Graphics for Chemical Structures

ACS SYMPOSIUM SERIES **341**

Graphics for Chemical Structures

Integration with Text and Data

Wendy A. Warr, EDITOR
Imperial Chemical Industries

Developed from a symposium sponsored by
the Division of Chemical Information
at the 192nd Meeting
of the American Chemical Society,
Anaheim, California,
September 7–12, 1986

American Chemical Society, Washington, DC 1987

Library of Congress Cataloging-in-Publication Data

Graphics for chemical structures.
  (ACS symposium series, ISSN 0097-6156; 341)

  Includes bibliographies and indexes.

  1. Chemical structure—Data processing—
Congresses. 2. Computer graphics—Congresses.

  I. Warr, Wendy A., 1945-   . II. American
Chemical Society. Division of Chemical Information.
III. American Chemical Society. Meeting (192nd: 1986:
Anaheim, Calif.) IV. Series.

QD471.G753   1987   541.2′2′028566   87-3575
ISBN 0-8412-1401-8

QD
471
.G753
1987

PRINTED IN THE UNITED STATES OF AMERICA

Foreword

The ACS SYMPOSIUM SERIES was founded in 1974 to provide a medium for publishing symposia quickly in book form. The format of the Series parallels that of the continuing ADVANCES IN CHEMISTRY SERIES except that, in order to save time, the papers are not typeset but are reproduced as they are submitted by the authors in camera-ready form. Papers are reviewed under the supervision of the Editors with the assistance of the Series Advisory Board and are selected to maintain the integrity of the symposia; however, verbatim reproductions of previously published papers are not accepted. Both reviews and reports of research are acceptable, because symposia may embrace both types of presentation.

Contents

Introduction

by Wendy A. Warr

Pharmaceuticals Division, Imperial Chemical Industries, Macclesfield, Cheshire SK10 4TG, England

THE WIDESPREAD USE of microcomputers has led to the development of software packages that integrate chemical structures with data and text. Many commercially available and in-house systems for handling chemical structures exist; some systems handle data to a greater or lesser extent. There are data base management systems for data and word processing packages for text. The way ahead lies in integration. Software packages that integrate chemical structures with data or text fall into four categories:

- scientific word processing packages,

- packages that allow graphics entry of chemical structures but do not allow substructure searching or interfacing with other systems,

- packages that allow graphics structure entry and substructure searching and interfacing with other systems, and

- packages designed to act as front ends to molecular modeling systems.

Software Packages

First, many scientific word processing packages exist. Some can handle chemical structures. Examples (not necessarily chemically oriented) are:

Advent Series 70 (Advent Systems, Barre, VT)
Chem Word (Laboratory Software, Aylesbury, England)
CPT Phoenix (CPT, Minneapolis, MN)
Z Egg (Peregrine Falcon Company, Sausalito, CA)
IS-Genius (Infosystems N.V., Vilvoorde, Belgium)
Mass-II (Microsystems Engineering Corporation, Hoffman Estates, IL)
Micro T_EX (Addison–Wesley Publishing, Reading, MA)
PCT_EX (Personal T_EX, Mill Valley, CA)
Proofwriter (Image Processing Systems, Madison, WI)
Protex/Scientex (Scientex, Stevenage, England)
Samna Word III (Samna Corporation, Atlanta, GA)
Spellbinder Scientific (Lexisoft, Davis, CA)

T^3 (TCI Software Research, Las Cruces, NM)
Techfont (Santa Barbara Technology, Santa Barbara, CA)
TechWriter (Computer Mart, Waltham, MA)
Volkswriter Scientific (Lifetree Software, Monterey, CA)
Vuwriter (Vuman Ltd., Manchester, England)
Word Mark (MARC Software International, Palo Alto, CA)

These packages can be divided into two classes: the WYSIWYG ("What You See Is What You Get") programs, which require the computer to operate in graphics mode at least part of the time, and mark-up language programs, which require the computer to operate only in text mode. All these programs are beyond the scope of this book (*1–4*).

The second category includes packages that allow graphics entry of chemical structures by using a mouse or something similar but do not allow substructure searching or interfacing with other systems. Molecular Presentation Graphics (MPG, Hawk Scientific Systems, Kinnelon, NJ) is a software package in this category. Unfortunately, an MPG user was not available to write a paper. The program was designed specifically for chemists; is easy to learn and use (it has function keys or a mouse for drawing structures); has structure storage and modification facilities; and allows merging of diagrams into text files. A similar program, DATALYST, stores data related to chemical structures. MPG, however, differs from most of the software discussed in this book—it does not store structures as connection tables. Therefore, substructure searching and other functions are not possible.

Another package in this category, Wisconsin Interactive Molecular Processor (WIMP, written by H. W. Whitlock, University of Wisconsin, Madison, WI and marketed by Aldrich Chemical Company, Milwaukee, WI [*5*]) permits structure entry and high-quality output (again without substructure searching or interfacing with other systems).

A final example in this category is ChemDraw (Stewart Rubenstein, Somerville, MA), which probably uses connection tables but does not have substructure searching or interfacing with other software. ChemDraw appears in this book (*see* Chapter 3) because it is probably the most user-friendly package on the market for entering chemical structures and producing high-quality output. Unfortunately, it is available for use only on Apple Macintosh computers.

More than half of the chapters in this book are heavily oriented toward the third category of software: that which allows graphics structure entry with a mouse or something similar and can store connection tables so that substructure searching and interfacing are possible in principle or in practice. Good examples are The Chemist's Personal Software Series (Molecular Design Limited, San Leandro, CA) and the PSIDOM family of programs (Hampden Data Services, Abingdon, Oxford, England). Software of this type is a fast-growing topic of considerable current importance. Very few software reviews or papers authored by users have appeared on this topic. This book covers all graphics for chemical structures that I knew of in 1986 except Graphic Input and Output of Structures (GIOS, written by Gunter von Kiedrowski and marketed by Gearg Thieme Verlag, Stuttgart, Federal Republic of Germany and Thieme, Inc., New York, NY).

Chemical structure entry packages that were designed specifically as front ends to molecular modeling systems constitute the fourth category. Examples are ChemCad (C Graph Software, Austin, TX); XI-CHEM-DRAW (Xiris Corporation, New Monmouth, NJ); and ChemNote (Polygen, Waltham, MA). In the future, some of these packages could compete with the third category of software, but strong evidence of competition does not exist now.

To cover the specialized area of chemical structures in detail, authors could provide only cursory coverage of mathematical expressions and tabular material. Many of the scientific word processing packages that I mentioned and some software the book discusses in detail (especially ChemText from Molecular Design Limited) will handle scientific and graphics material other than chemical structures. This book does not cover well-known graphics systems that handle chemical structures or chemical reactions; these systems are already well-documented (6, 7).

Compact Disk–Read-Only Memory

No timely book on the integration of graphics and text would be complete without a mention of compact disk–read-only memory (CD–ROM). This 4.72-inch optical disk (much like the compact disks used for high-fidelity audio) holds digital data (coded text, digitized images, vector graphics, and software). CD–ROM's enormous storage capacity (about 600 megabytes) is particularly useful for graphics, which take up more space than text. I know of only one CD–ROM product that handles chemical structures specifically: the Hampden Data Services/Pergamon–Infoline/KnowledgeSet joint venture mentioned in Chapter 2.

As an electronic publishing medium, CD–ROM can store an unlimited number of fonts; frequently used graphics images; and entire document pages (both text and graphics) from catalogs, directories, and manuals. When used to search in-house bibliographic data bases, CD–ROM is best suited to information that does not become outdated rapidly and that is accessed on a regular basis. The time delays and costs involved in mastering and replicating new versions of a data base mean that the information can never be current. However, in-house users can search as often as they like and take as long as they wish formulating and running searches because there are no additional costs once CD–ROM has been purchased. Downloading and processing the results may also be possible without additional cost.

Making a master CD-ROM disk is expensive (about $2000); replicates are cheap (about $10). The enormous success of compact disk audio players, the similarity of compact disks to CD-ROM players, and the growth of the microcomputer industry have opened up the market. Eventually, economies of scale could make the CD-ROM inexpensive enough for widespread use.

Summary of Chapters

In a book of this type, the various chapters will inevitably be rather disconnected. This section is intended to put them into context. Chapter 1 is an overview of graphics for chemical information and Chapter 2 reviews the impact of microcomputers. Chapter 3 presents the point of view of a chemist in the pharmaceutical industry. Chapter 4 compares microcomputer software packages that manage chemical structures. In Chapters 5 and 6, the authors discuss two of these microcomputer software packages in detail. Chapter 7 describes a software package not covered in Chapter 4. Chapter 8 concerns a structure entry package that helps with use of the Beilstein Handbook. Chapter 9 looks at a major system, Beilstein Online. Perhaps this system should have been excluded, as were CAS ONLINE, DARC, and other large systems, but this chapter fits in well as a follow-up to Chapter 8. Chapter 10 is also concerned with a very large data base, but the graphics aspects are relevant and interesting. Chapters 9 and 10 involve the handling of large quantities of data related to chemical structures. Chapter 11 describes a chemical word processing system specifically for use on a mainframe machine (or "super mini"). As mentioned previously, most software of this type is written for microcomputers. Chapter 12 contributes technically oriented information on graphics. This area is interesting and significant in its own right but also is important because of the preeminent position of Chemical Abstracts Service in the field of chemical structure data bases. Chapter 13 covers a completely different aspect of graphics: storing and retrieving diagrams in patents by using vectorized graphics.

Acknowledgments

I thank the officers and program committee of the ACS Division of Chemical Information and the authors who contributed to this book.

References

1. *See* Marshall, J. C. *J. Chem. Inf. Comput. Sci.* **1986,** *26,* 87–92.
2. *See Anal. Chem.* **1985,** *57(8),* 888–892.
3. *See* Cullis, R. *Pract. Comput.* March, **1986,** 96–97.
4. *See* Einon, G. *PC User* July, **1985,** 105-115; August, **1985,** 149-154.
5. Reviewed in *Comput. Appl. Biosci.* **1986,** *2(4),* 333–335.

6. *Communication, Storage and Retrieval of Chemical Information;* Ash, Janet; Chubb, Pamela; Ward, Sandra; Welford, Stephen; Willett, Peter, Eds.; Ellis Horwood: Chichester, England, 1985.
7. *Modern Approaches to Chemical Reaction Searching;* Willett, Peter, Ed.; Gower: Aldershot, England, 1986.

March 2, 1987

Chapter 1

Evolution of Molecular Graphics

W. Todd Wipke

Department of Chemistry, University of California, Santa Cruz, CA 95064

Graphics have played an important role in the development of chemistry and computer assistance in chemical research has been intimately intertwined with computer graphics. This paper reviews the impact of computer graphics in chemist-computer communication and representation of chemical information.

The development of computer graphics and its impact on chemistry covers over twenty years and many publications. The objective of this paper is to highlight what in my opinion are key features of this field. I have constrained the scope of the discussion to the sub-area of interactive graphics, excluding the application to molecular modeling. While molecular modeling applications of graphics are extremely important today and have received extensive attention in both the scientific and popular literature,(1-4), they represent only a small fraction of graphic applications in chemistry. Thus this paper will focus on the use of computer graphics as the medium for bidirectional communication between chemist and computer and between chemists via computer.

The natural medium for chemists, particularly organic chemists, is the graphical language of the structural diagram, the invention of which greatly accelerated the development of organic chemistry and the theory of chemical structure. Over time chemists have developed graphical notations for types of bonding, stereochemical configuration, and of course the connectivity of the molecule. The structural diagram is as close to the essence of classical chemical structure as we can currently get, consequently, chemists think in terms of structural diagrams and their perception of same is direct and rapid. Linear encodings of the structural diagram such as Wisswesser Line Notation

L E5 B666TJ

Wiswesser Line Notation

or systematic IUPAC chemical nomenclature, while interconvertible via algorithm to the diagram, are not as easily perceived by the chemist because the connectivity of the molecule is implicit via rules rather than explicit via graph. Reading or writing linear notations takes additional mental processing which can lead to errors. Thus for the chemist, it was clearly advantageous to communicate with the computer and with colleagues in natural graphical language. But for the computer system developer, the linear notation was more attractive owing to the ease of reading, writing, and comparing character strings, the simplicity of data structures for strings, the lower cost of text-only devices, the fascination for coding systems in general, and sometimes the desire to be an information gatekeeper to translate the chemist's questions and the system's responses. Computer memory and processor time in the early days of computers were too valuable to spend on the more complex algorithms needed to cope with graphical communication.

Graphical Input of Chemical Structures

The first graphical entry of a chemical structure was reported by Opler using a light pen for drawing(5). Light pens are passive devices, i.e., they only interrupt the computer when they see light. The computer program has to know where it was painting on the screen at that time, compute the coordinates, and reposition the "tracking cross" centered on the new location. A micro switch in the pen reports if the pen is pressed against the screen. Quick motions of the pen by the user often cause the cross to be left behind, causing the same frustration as a ball point pen that is running out of ink. Since the screen is normally vertical, the light pen requires holding one's arm up for extended periods of time.

The OCSS-LHASA (Organic Chemical Synthesis Simulation-Logic and Heuristics Applied to Synthetic Analysis) synthesis planning system developed by Corey and Wipke(6) was the first system to allow drawing the chemical structure using a "Rand Tablet." The technology spread to the NIH Prophet system and Feldmann's substructure search system(7). The tablet itself had a grid of wires and the pen was an antenna that picked up the signal from the closest wires. In contrast to the light pen, the tablet was active and directly reported to the program the absolute coordinates of the current pen position and the pen status. The user no longer had to drag a tracking cross around, but could point anywhere. A micro switch in the pen reported whether the pen was being depressed hard, and a proximity circuit told if the pen was on the surface of the tablet or within an inch of the tablet. Since the tablet is horizontal, the hand rests naturally as on a writing pad. Most users quickly accommodate to writing on the tablet but looking at the screen, because the coordinate system of the tablet is easily mentally mapped onto the coordinate system of the screen. Right on the tablet is right on the screen, up on the tablet is up on the screen.

The three-dimensional acoustic tablet we developed in 1970 (see Figure 1) enabled one not only to draw in two dimensions, but also in three dimensions(8). The tablet consists of three orthogonal strip microphones, the pen emits a spark, and the tablet measures the time required for the sound of the spark to reach the microphones, giving the perpendicular distance to the three microphones. Simple equations relate these distances to the Cartesian coordinates to within 0.01 inch in a 14 inch cube. When the pen was lifted off the surface more than an inch, the coordinate system of the tablet was changed so that the 3-D tablet cube space corresponded exactly to the 3-D display space. The SECS (Simulation and Evaluation of Chemical Synthesis) program, a three-dimensional synthesis planning program, used this system in conjunction with the Evans and Sutherland LDS-1 display connected to a PDP-10 computer(9). By displaying the image in stereo, and the tracking cross in stereo, one was permitted to trace a three-dimensional Fieser model, or to reach in and move atoms interactively in three dimensions, even while an energy minimization process was in progress. A finger operated pressure switch on the side of the pen allowed the user to designate "pen-down" status when pointing to a position within the 14 inch "cubic drawing space".

Both the Harvard PDP-1 DEC-340 display system, and the Princeton PDP-10 E&S LDS-1 were unique and expensive. The appearance of the DEC GT40 terminal which provided vector drawing display, PDP-11/05 minicomputer with 16 K bytes of memory, and light pen in 1973 for $15,000 opened the way for more widespread use of graphical input(9). SECS and LHASA were both adapted to the GT40 and the MACCS system(10-12) initially used it also. MACCS is a Molecular Access System for interactively storing and retrieving chemical structures and related data, first described by Marson, S. A.; Peacock, S. C.; Dill, J. D.; Wipke, W. T., "Computer-Aided Design of Organic Molecules", at the 177th National Meeting of the American Chemical Society, Honolulu, Hawaii, April 1979; and later by Wipke, W. T.; Dill, J. D.; Peacock, S.; Hounshell, D., "Search and Retrieval Using an Automated Molecular Access System", at the 182nd National Meeting of the American Chemical Society, New York, NY, August 1981. The IMLAC terminal then reigned for several years after the GT40 was discontinued until the DEC VT640 raster display with light pen became the low cost replacement. The DEC VT640 terminal had the advantage over the TEK 4010 storage tube terminal in that the VT640 could selectively erase a line without clearing the entire screen.

Today we find the IBM PC or Apple Macintosh used as VT640 or TEK 4010 emulators with the mouse as the preferred drawing device. The mouse is "active" like the tablet, but in contrast to the tablet which provides absolute coordinates, the mouse provides only relative coordinates and the mouse can not be used to trace a hard copy diagram like a tablet can.

In twenty years graphical input has come from isolated research laboratories to be implemented in commercial personal computer products.(13,14) The graphical display technology switched from expensive analogue vector displays to cheaper "TV-type" raster technology, pointing devices became cheap from mass production, and the cost of memory, processor, and disk dropped while power increased. The Harvard PDP-1 computer which occupied an entire room had only 48K bytes of memory and a 500 K word magnetic drum for storage. The IBM XT provides 640K bytes of memory and 10 Mbyte of disk storage.

Optical Recognition. Off-line entry of hard copy structural diagrams was developed by Meyer(15) using an optical reader that could recognize lines drawn on a special paper grid. The paper tape that was produced was later processed by computer into a connection table. Today there are optical scanners to convert hand drawn engineering drawings into line and text entities for use by CAD/CAM programs, but the methodology has not been utilized in a practical chemical entry system, partially because interactive drawing is easier, and because of problems from interpreting crossed bonds, stereochemistry, etc.

Representation of Molecular Structure

There are many mathematically equivalent methods of representing the classical structure of a molecule, e.g, adjacency matrix, connection table, or linked list. The OCSS program was the first to represent bonds as named entities, rather than just the thing between two named atoms.(9) Besides the chemical perception reasons for this choice, it simplified drawing the molecule on the screen so each bond was drawn only once.

Valid Stereochemical Notations

Stereochemistry. Algorithms have been developed for perception of any valid projections of a tetrahedral stereocenter and, given the configuration, to display a valid stereochemical notation for that center(16). Higher coordination stereochemistry has also been perceived graphically by algorithm(17,18). Unfortunately, chemists have different views about what the wedging and hashing of bonds means. Everyone agrees that the wedge means the substituent is *up*, and the hash means the substituent is *down*, but since there is no *up* or *down* in the molecule, the differences arise in choice of the *reference plane*. Some choose the plane of the paper, others choose the approximate molecular plane which is sometimes perpendicular to the plane of the paper, while others choose a local reference plane formed by the bonds that are not marked. Further sometimes chemists are not consistent in their notation and even worse, they mix perspective three-dimensional portrayal with two dimensional configuration notation. The unfortunate fact is that in the realm of stereochemistry, chemists sometimes have difficulty communicating unambiguously with each other. The epoxide ring in Figure 2, taken from a published paper, has stereochemical notation that can not be unambiguously interpreted.

Conformation Representation. Most work to date has been concerned in conveying to a computer the structure of the molecule with stereochemical configurations, but not the conformation of the molecule. For approximate input to Allinger's MM2 program, a popular molecular mechanics calculation program, drawing programs have been built that allow one to indicate on which level behind or in front of the plane of the paper each atom is located, generally by relative "up", "down" commands during the drawing. It remains a challenge to devise a program that can properly perceive a perspective drawing of a particular molecular conformation, even assuming a reasonably accurate drawing such as the one shown below. Such a program would facilitate entry to molecular modeling programs.

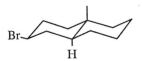

Molecular Shape Representation

Extensive work has been directed toward computer presentation of space-filling models, and molecular surfaces, but there had not been occasion to draw shapes until the need arose to search for pharmacophoric patterns, i.e., arrangements of atoms in three dimensions relative to each other(19-21). We can expect to see continued work in shape description as systematic drug design efforts increase. It seems entirely reasonable that the chemist may wish to ask for a molecule that is more "heart-shaped" to fit a given receptor. But is that heart as in "human", or heart as in "Valentine"? Defining a graphical input notation for shape that the computer can understand remains for now an interesting problem.

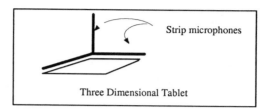

Figure 1.   The three-dimensional acoustic tablet and spark pen.

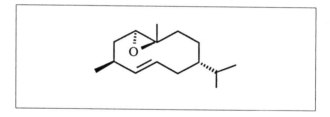

Figure 2.   An example of published ambiguous stereochemical notation (the epoxide).

Chemist to Chemist Communication

Molecular graphics are also used as we mentioned earlier as the natural language for communication between chemists through seminars and printed publications. Making projection slides and transparencies for scientific presentations previously required hours of careful drawing with pen and ink, or working with press on stencils. Figures for printed publications are similarly time consuming and expensive to create, and generally do not make good slides for oral presentations. The introduction of computer-aided drafting software such as AutoCad (AutoDesk, Inc.) on the IBM PC; and MacDraw, MacPaint (Apple Inc.), ChemDraw (Rubenstein) on the Macintosh together with high resolution plotters and the Apple LaserWriter has provided an alternative method of creating high quality presentation molecular graphics. Each of these programs has limitations, either in the drawing of chemical structures or the handling of text, and none of them really integrate text and graphics.

A new frontier in chemist-chemist communication is opened by ChemText (Molecular Design Limited), for IBM PC compatible computers (see D. d. Rey, "Applications of the Chemists Personal Software Series in the Chemists Workstation", this volume). Graphics and text are given equal status and molecular graphic images are *computable. Computable image* means that when an object such as a molecule or reaction is inserted as an image, that there is no information loss-- the object can later be extracted from the document. Thus a chemist receiving an electronic ChemText document can extract the molecules, register them in a data base or submit them to MM2 for energy calculations. This clearly gives more meaning to the molecular graphics in the document and facilitates scientific collaboration through more efficient chemist-chemist communication.

Computer to Computer Communication

Finally, I would like to briefly describe developments in communications *between* computers for molecular graphics purposes. There are several key points, the first of which is the transfer of molecule representations between different programs and between personal computers and mainframes. The recent development of standards for molecular connection tables, ranging from the Brookhaven Protein Data Bank format to the Molecular Design Limited MOLFILE format has eliminated much of the Tower of Babel, and permitted chemists to utilize programs from different sources on different machines to solve the problem at hand without writing conversion programs.

A second important advance is the emergence of partial standards in graphic terminal protocols, particularly for lower performance terminals. The VT100 retrographics, and the Tektronix 4100 series have become default standards. But at the high performance end of the scale, three-dimensional graphic standards are still unresolved. It is very difficult to handle a Silicon Graphics IRIS terminal the same as an Evans and Sutherland PS300, because the internal concepts are quite different, and one is raster, the other is vector. Fortunately, graphic program developers now generally support a wide variety of graphic devices in a way transparent to the chemist user.

The last advance to be mentioned is networks. Standards such as TCP/IP have enabled different computers to communicate with each other at high speed via

Ethernet. But at the application level, decisions still have to be made as to which tasks to assign to which computer on the network in so-called *distributed computing*. Distributed computing raises many questions about how much a workstation should do versus the mainframe, and molecular graphics is one of the computationally intensive tasks that needs to be allocated. We can look forward to some interesting studies and systems in this arena in the future.

Conclusion

Thus computer graphics impacts the chemist in two major ways. First it simplifies access to computer stores of chemical information and to chemical computations for discovering new information. Second, it facilitates transmission of chemical information via publication quality graphic-containing documents for dissemination in oral presentations, in hard copy publications, or in electronic mail; and via computer networks. Graphic images which are also *computable* give new power to molecular graphics and the documents that contain them.

Acknowledgments

The author thanks Dr. Warr for the invitation to participate in this volume. This paper was prepared with ChemText and printed on an Apple LaserWriter.

Literature Cited

1. Levinthal, C. *Sci. Am.* **1965**, *214*, 42.
2. Marshall, G. R.; Bosshard, H. E.; Ellis, R. A. In *Computer Representation and Manipulation of Chemical Information;* Wipke, W. T.; Heller, S. R.; Feldmann, R. J.; Hyde, E., Eds.; John Wiley and Sons, Inc.: New York, 1974; pp 203-238.
3. Gund, P.; Andose, J. D.; Rhodes, J. B.; Smith, G. S. *Science* **1980**, *208*, 1425-1431.
4. Langridge, R.; Ferrin, T. E.; Kuntz, I. D.; Connolly, M. L. *Science* **1981**, *211*, 666.
5. Opler, A.; Baird, N. *Am. Doc.* **1959**, *10*, 59-63.
6. Corey, E. J.; Wipke, W. T. *Science* **1969**, *166*, 178.
7. Feldmann, R. J.; Heller, S. R. *J. Chem. Doc.* **1972**, *12*, 48-54.
8. Wipke, W. T.; Whetstone, A. *Computer Graphics* **1971**, *5*, 10.
9. Wipke, W. T. In *Computer Representation and Manipulation of Chemical Information;* Wipke, W. T.; Heller, S. R.; Feldmann, R. J.; Hyde, E., Eds.; John Wiley and Sons, Inc.: New York, 1974; pp 147-174.
10. *Chem. Eng. News*, June 18, **1979**, 29.
11. Adamson, G. W.; Bird, J. M., Palmer, G., Warr, W. A. *J. Chem. Inf. Comput. Sci.* **1985**, *25*, 90-92.
12. Anderson, S. *J. Mol. Graphics*, **1984**, *2*, 83-90.
13. Curry-Koenig, B.; Seiter, C. *Am. Lab.* **1986**, *18, (5)*, 70-78.
14. Seither, C.; Cohen, P. *Am. Lab.* **1986**, *18, (9)*, 40-47.
15. Meyer, E. In *Computer Representation and Manipulation of Chemical Information;* Wipke, W. T.; Heller, S. R.; Feldmann, R. J.; Hyde, E., Eds.; John Wiley: New York, 1974; pp 105-122.
16. Wipke, W. T.; Dyott, T. M. *J. Am. Chem. Soc.* **1974**, *96*, 4825, 4834.
17. Choplin, F.; Marc, R.; Kaufmann, G.; Wipke, W. T. *J. Chem. Inf. Comput. Sci.* **1978**, *18*, 110.
18. Choplin, F.; Dorschner, R.: Kaufmann, G.; Wipke, W. T. *J. Organometallic Chem.* **1978**, *152*, 101.

19. Gund, P.; Wipke, W. T.; Langridge, R. In *Computers in Chemical Research and Education;* Elsevier: Amsterdam, 1973; Vol. II, pp 5/33-38.
20. Gund, P. *Ann. Repts. Med. Chem.* **1979**, *14,* 299-308.
21. Wipke, W. T. Abstracts of the 186nd National Meeting of the American Chemical Society, Washington, DC, August 1983.

RECEIVED March 25, 1986

Chapter 2

Combining Chemical Structures with Data and Text on a Microcomputer

William G. Town

Hampden Data Services, Hampden Cottage, Abingdon Road, Clifton Hampden, Abingdon, Oxon OX14 3EG, England

Chemists need to be able to combine chemical structure diagrams with text to produce reports, scientific papers, and safety and marketing material. They also need to be able to conduct searches which combine structural concepts with text and/or data concepts and they require microcomputer systems which would replace the traditional index card file. Ideally, the personal chemical information manager would provide off-line query negotiation and uploading and would also allow downloading of both chemical and textual information from public and company files into the desktop system as well as providing access to CD-ROM based products. The extent to which these concepts are realized by existing software is analyzed and projections for the future are made.

Since the early days of computing, organizations concerned with chemical substances have sought to make chemical structure information searchable and displayable by computer. The first of these systems were batch systems depending on modified terminals and/or special print chains for their graphical input and output capabilities. Examples of these early systems include the ICI CROSSBOW system (1), which used a modified print chain to output chemical structure diagrams generated from a Wiswesser Line Notation and the Walter Reed Army Institute of Research's Chemical and Biological Information Retrieval System (2) which used a chemical typewriter for structure input and a modified drum printer for structure output. In a novel approach BASF used a photo-electric scanner for input of chemical structures to the GREMAS system (3). Chemical Abstracts Service pioneered the production of high quality structure diagrams in the 1960s and soon integrated their photocomposed structure diagrams into the production cycle of Chemical Abstracts (4). These few examples illustrate the multitude of developments in this field during the 1960s.

0097-6156/87/0341-0009$06.00/0
© 1987 American Chemical Society

        Progress in chemical structure handling systems has closely
followed developments in computers in general and operating systems
in particular. Early in the 1970s development of the first
interactive chemical information systems followed the availability of
commercial time-sharing computer systems. In particular, working at
the Division of Computer Research and Technology (DCRT) of the
National Institutes of Health at Bethesda, Maryland, Feldmann, Heller
and others developed the DCRT Chemical Information System (5) which
subsequently became the widely known and used NIH/EPA Chemical
Information System (CIS, 6). Although the early developments of
CIS were made using graphical terminals, by 1973 a technique for
inputting and visualizing chemical structure queries on teletype
(TTY) compatible terminals had been developed (7). This technique
was a key factor in the wide acceptance of CIS as it made chemical
graphics capabilities, however crude by today's standards, available
to chemists using low-cost TTY terminals.
        A second generation of interactive chemical information systems
began to appear in the early 1980s using graphics devices such as the
Hewlett Packard 2647A or Tektronix 4010 terminals. These lower cost
(10 - 15000 dollar) graphics terminals offered the possibility of
higher quality chemical structure diagrams. Public on-line services
using these facilities quickly followed: CAS ONLINE in 1980 (8) and
QUESTEL - DARC in 1981 (9). Initially, in CAS ONLINE, substructure
search queries were formulated using screen numbers from a dictionary
containing almost 6000 different structural features. However, late
in 1981 the ability to describe a structure query by drawing the
complete or partial molecule on the graphics terminal became
available. Even today, however, both CAS and Telesystemes Questel
offer TTY input techniques as well as graphical input of queries.
The advent of personal computers and terminal emulation programs has
reduced the cost of high quality graphics displays by almost an order
of magnitude and probably the majority of users of these public
systems now use this type of equipment.
        Development of software for in-house chemical structure
information systems has paralleled the development of public systems.
One of the earliest commercially available systems for in-house use
was the CROSSBOW system mentioned above. Initially the software was
distributed by the developer, ICI Pharmaceuticals Division, but later
Fraser Williams (Scientific Systems) Ltd assumed responsibility for
marketing. This batch system was installed in some 25 locations and
dominated the market until the late 1970s.
        Two systems using graphics input and output emerged in the late
1970s: the DARC system, developed by Dubois and co-workers at the
University of Paris and later marketed by Télésystèmes Questel, and
the MACCS system, which was developed and marketed by Molecular
Design Ltd. The latter system has dominated the in-house market
during the early 1980s with over 120 installations worldwide,
although more recently Johnson and co-workers at the University of
Leeds are challenging this domination with ORAC and OSAC
(respectively for reaction and substructure searching) and Fraser
Williams is promoting SABRE (a software package for the registration
and search of chemical structures based on connection tables or
Wiswesser Line Notation).

Impact of microcomputers on chemical structure information systems

Just as earlier developments in computer systems have influenced the
direction of development of chemical structure information systems in
the past, a new generation of systems now emerging as a result of the
microcomputer revolution is causing a change in direction and
emphasis. Although the early microcomputers were regarded by many as
"toys", the entry of IBM into personal computers in 1982 has so
revolutionized the market that few successful microcomputer
manufacturers now market computers which are not "IBM compatible" and
few large companies will buy non IBM compatible machines.
Microcomputers are now so powerful that alone they perform many of
the tasks which until recently would have required a mainframe
computer. However, their potential for linking into networks with
other microcomputers and/or mainframe computers greatly expands the
range of applications. As mentioned above, in the domain of chemical
structure handling systems, microcomputers have already influenced
the development of on-line services by making high quality graphics
more cheaply available. However, the revolution is just beginning
and microcomputer graphics will impact on many areas of chemical
activity, including: chemical report production, preparation of
visual aids, personal chemical data bases, off-line query negotiation
and uploading, downloading of chemical structures, graphical "front
ends" to many types of chemically oriented software, and, using the
increasingly flexible distributed processing systems now available,
integration of these functions with the more traditional centralized
chemical information retrieval systems will be possible.

Chemical report production

The inclusion of chemical diagrams in typed reports has caused
problems since the typewriter was invented. The traditional approach
was to "cut and paste" hand-drawn or stenciled diagrams into the
typed text. This approach was carried over into the early word
processing systems but here there were two disadvantages. Firstly,
the chemical diagram was not part of the electronic record of a
document and therefore not included in any electronic archive.
Secondly, the advent of word processing facilitated the production of
multiple revisions of a document but, when chemical diagrams were
included, careful copying of the diagram to each generation of the
document was required and, implicitly, careful checking of the
resulting collage was necessary.
    This problem was recognized from the early days of computing and
the chemical typewriters of the 1960s represented one approach to its
solution which, in a modified form, is still in use today. Chemical
structures can be displayed using a character matrix approach in
which each character has a fixed width and occupies a cell in a fixed
grid or matrix (not visible). To depict chemical structures
recognizably by this technique requires the introduction of
additional characters which may be limited to forward and backward
sloping lines (Figure 1) or may additionally include line
intersections for ring vertices and fusion points (Figure 2). Both
dedicated word processors and microcomputer word processor software
packages using this technique are still being marketed today. The

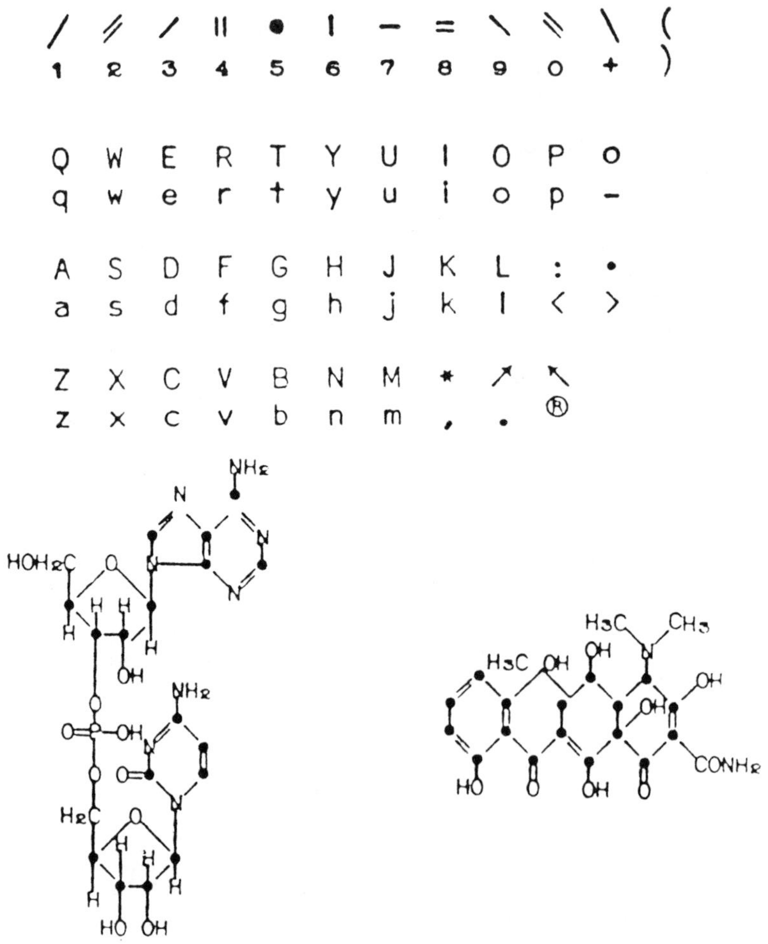

Figure 1.　Character set for the Dura chemical typewriter and sample chemical structure output.

limitations of this approach are, firstly, that the angles available
for drawing bonds are restricted, which makes some bridge ring
structures difficult to draw, and, secondly, that the imperfections
of the output devices are sufficiently large so as to make the joins
between the various segments of a line visible (Figure 3). However,
the approach is used successfully by at least one manufacturer of
photocomposers where the high resolution and large character set used
produces pleasing results. Current software packages based on this
approach include T3 (TCI Software Research), The Egg (Peregrine
Falcon), Vuwriter (Vuman Computer Systems), Spellbinder (Lexisoft)
and ChemWord (Laboratory Software).

The second major group of software products aimed at this
market are either Computer-Aided Design (CAD) packages or other
drawing programs more or less adapted to the requirements of
chemistry. The unmodified CAD packages such as AutoCAD (Autodesk)
and Doodle (Trilex) make no concessions to chemistry and although the
results can look good, these packages are extremely tedious to use
for this purpose. Other packages such as ChemDraw (Stewart
Rubenstein), WIMP (Wisconsin Interactive Molecule Processor from the
University of Wisconsin) and MPG (Molecular Presentation Graphics
from Hawk Scientific Systems) include such features as predrawn
structural templates which are tailored to chemical diagram
production. However, they do not include the minimal chemical
knowledge required for valency checking and are therefore dangerous
in the hands of the non-chemist. None of these systems produces a
connection table which can be used for other purposes and although
MPG and ChemDraw have been successful in the IBM PC and Apple
Macintosh market sectors respectively, they are likely to be
overtaken by more general and sophisticated packages such as the
Chemists' Personal Software Series (CPSS from Molecular Design Ltd)
and PSIDOM (Professional Structure Image Database on Microcomputers
from Hampden Data Services).

Personal chemical data bases

The chemical structure retrieval software described earlier has given
the owners of public or in-house chemical information systems the
possibility to offer to their customers facilities for searching by
structure and substructure. The new generation of microcomputer
software is now offering similar facilities to small companies,
project teams and individuals. The ability to represent structural
queries graphically and to review the results of the search using
high quality graphic displays is now available at a cost within the
individual's budget. In addition the diagrams stored in files
generated by microcomputer packages such as CPSS or PSIDOM can be
incorporated into reports or used for high quality diagram production
(Figure 4).

The chemical structure files built by these software packages
consist of connection tables, a form of structural representation
used in the CAS Chemical Registry System, and in consequence by both
CAS ONLINE and DARC, and incidentally in many other systems
(including MACCS). Both CPSS and PSIDOM include modules for building
chemical structure data bases on personal computers which, in the
case of PSIDOM, are limited only by the capacity of the disk storage

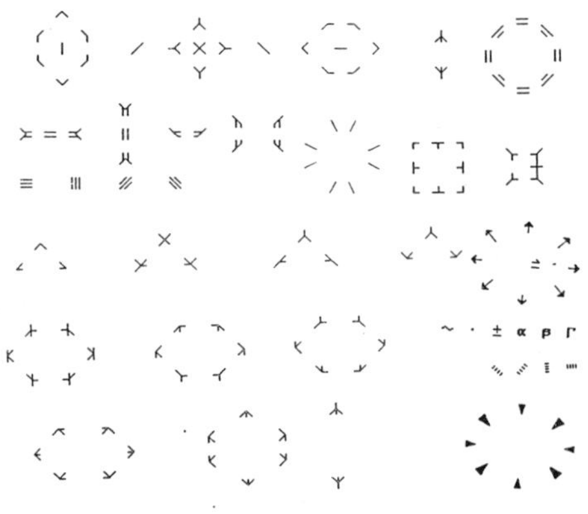

Figure 2.   ChemWord character set.

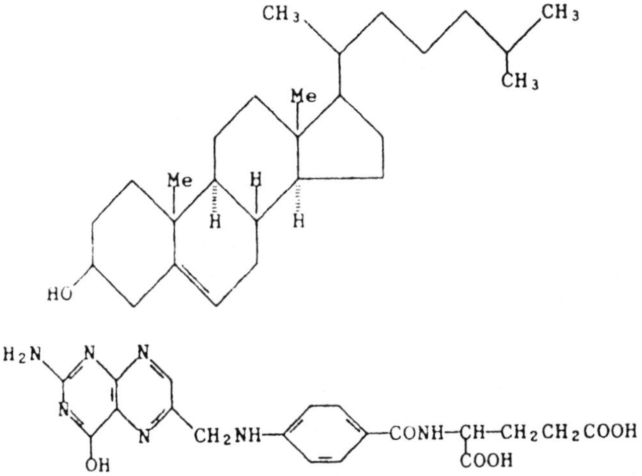

Figure 3.   Examples of structures typed with the EXXON 500/965
Chemical Editor.

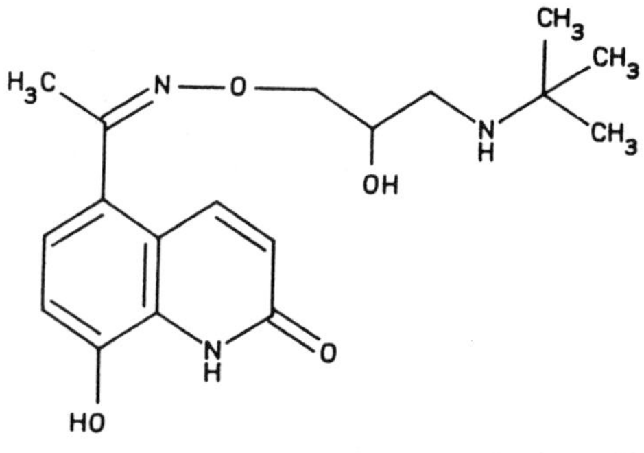

Figure 4.   A PSIDOM structure output on a Hewlett Packard plotter.

available on the microcomputer. Files of over 5000 structures have already been created using PsiGen, the drawing module of PSIDOM.

An alternative approach to creating a personal data base is to buy a ready-made data base. The Institute for Scientific Information (ISI) has already entered into agreements with Molecular Design Ltd and Academic Press (who market a low cost structure search package called ChemSmart) to distribute segments of ISI's Index Chemicus data base in the required form for each package. ISI has indicated a willingness to make its data base available in other formats and it is expected that PSIDOM versions will appear in 1987.

The recent development of the CD-ROM has opened up the possibility of distributing data bases on optical disks as well as on magnetic media by adapting the laser disk technology used in audio compact disks for use with personal computers. The differences between the two media are sufficiently great as to warrant a brief parenthetical comment. The first major difference is the recording and address mode of each medium: whereas floppy and hard disks are logically arranged in "cylinders" and records are indexed by their surface, track and sector addresses, on the CD-ROM data are recorded in a long spiral track which covers the whole surface of the disk and data are indexed by the playing time from the start of the disk. In other words, CD-ROMs are more like the conventional gramophone (phonograph) record the organization of which was mimicked by the audio compact disk. CD-ROM players also differ from other disk drives in that the player maintains a constant linear velocity rather than a constant angular velocity (that is, the speed of revolution increases as the reading-head moves from the periphery to the center of the disk). These differences coupled with the large seek times (of the order of seconds rather than milliseconds) and the large data base size increase the complexity of software development for the CD-ROM.

At present optical disks are "read-only" but it is expected that erasable optical disks will appear in 1987. The characteristics of the CD-ROM make it an ideal medium for machine-readable publication. The capacity of a compact disk when used to store digital data is about 550 million bytes (about 500 times the capacity of a similar sized magnetic disk) which is sufficient to store the entire contents of a encyclopedia on one disk. Indeed, among the first commercial CD-ROM products was the Grolier Encyclopedia which consists of 26 printed volumes. Many of the initial products are reference works or handbooks for which the medium is ideal. Subsets of on-line data bases have been produced on CD-ROM but the relatively high cost of producing a master disk for a CD-ROM precludes frequent updating and products with this type of requirement are therefore not well-suited to the medium. Most products announced so far are textual and have no chemical relevance but one publisher, Pergamon, has announced a joint venture with Hampden Data Services and KnowledgeSet to develop a software package combining text- and structure-searching of a chemical data base published on CD-ROM. The first prototype is expected to be available in the Spring of 1987.

Microcomputer-based "front ends" to on-line systems

A "front end" is a computer software package, normally resident in a
microcomputer, which isolates the user of a public on-line service
from some of the complexities of the retrieval mechanism. At the
simplest level, it automates the process of accessing the host
computer from which the service is given. Steps in this process may
include: dialing the computer either directly or through a
telecommunications network, supplying the user's identifier and
password when prompted by the host computer and possibly selecting
the desired file or service for the user.

Some front end packages allow the user to prepare queries in
advance of the on-line session, either using a simple text editor or
by means of a menu-driven interface, to store the queries in a file
and to "upload" the query to the on-line host computer either
automatically after logging on or under the control of the user.
Clearly this approach leads to savings in both telecommunication
charges and "connect-time" based charges for use of the host
computer. At present (January 1987) neither CAS ONLINE nor DARC
QUESTEL allow uploading of structure queries from a microcomputer.
Given the frequently long time required to develop a complex
substructure search query in both these systems and the large amount
of telecommunications traffic generated by graphical query
definition, it would be highly desirable to extend the scope of
packages such as CPSS and PSIDOM to include such a facility. This
would benefit both the user and the host since many casual "end
users" (chemists) are deterred by the so-called "taxi meter
syndrome", i.e., the awareness that large charges are accruing while
they grope their way through the maze of the present graphical query
definition procedures.

In addition to uploading of queries, text-based front ends also
permit "downloading" of search results. The "downloaded" file may be
a verbatim transcript of a search session (or part of it) or may have
been subject to automatic editing, reformatting and/or merging
processes. Here the chemist is better served since terminal emulator
software such as PC-PLOT (Microplot Systems) or EMU-TEK (FTG Data
Systems) does at least provide for the capture of image files (as
plot instructions or vectors) for later display and printing or
plotting. However, neither CAS ONLINE nor DARC allows the
downloading of chemical structures in the form of connection tables.
Such a facility would permit the automatic capture of chemical
structure records into personal chemical structure data bases (or
"structure downloading"). Once a structure has been downloaded, the
chemist would be able to use the full capabilities of packages such
as CPSS or PSIDOM for further searches (text-, data- or
structure-based) or for combining structures with data and text for
data base construction or report production.

Another major characteristic of front end systems briefly
mentioned in the context of query uploading is the ability to perform
query translation. In general, each host computer uses its own own
command language and, particularly in the field of bibliographic and
other textual data bases, a query translator is desirable to ensure
that a particular query can be searched against the many and varied
relevant files on diverse host computers. In chemical structure
systems, the only query translator available to date is TopFrag

(Derwent) in so far as it generates a textual query file for patent searching from a chemical structure diagram. The number of computer systems which a chemist may be required to use is growing: in addition to on-line structure search systems he may be faced with separate in-house systems for structures and reactions, modeling systems, synthesis planning systems, spectral library search software and soon CD-ROM based handbooks each of which may have requirements for structure drawing and searching. The situation is becoming intolerable, the ideal solution being a universal graphical structure query builder which would allow searching in personal, company and public structure data bases and input to other chemical systems. Such a universal front end is a design goal of the PSIDOM package.

Literature Cited

1. Thomson, L.H.; Hyde, E.; Matthews, F.W. J. Chem. Doc. 1967, 7(4), 204-209.
2. Feldman, A.; Holland, D.B.; Jacobus, D.P. J. Chem. Doc. 1963, 3(4), 187-190.
3. Meyer, E. Angew. Chem. Int. Ed. Engl. 1965, 4, 347-352.
4. Kuney, J.H.; Lazorchak, B.G. Proc. Am. Doc. Inst. 1964, 1, 303-305.
5. Feldmann, R.J.; Heller, S.R.; Shapiro, K.P.; Heller, R.S. J. Chem. Doc. 1972, 12(1), 41-47.
6. Heller, S.R.; Milne, G.W.A.; Feldmann, R.J. Science 1977, 195, 253-259.
7. Feldmann, R.J. In Computer Representation and Manipulation of Chemical Information; Wipke, W.T.; Heller, S.R.; Feldmann, R.J.; Hyde, E., Eds.; John Wiley & Sons: New York, 1977; pp 55-81.
8. Farmer, N.A.; O'Hara, M.P. Database 1980, 3(4), 10-25.
9. Attias, R. J. Chem. Inf. Comput. Sci. 1983, 23(3), 102-108.

RECEIVED March 11, 1986

Chapter 3

Chemical Graphics: Bringing Chemists into the Picture

Trisha M. Johns

G. D. Searle & Co., Skokie, IL 60077

It is interesting to note that the literature of
chemical documentation through the past twenty years
has had a few central themes: one is the need for
systems to denote chemical structures in computer-
readable form, and another is the need to educate
chemists in literature retrieval. Until recently, one
theme has precluded the other: as long as the standard
computer systems were limited to non-graphical
representations, chemical information retrieval would
remain sufficiently mysterious to the end user chemist.
As structure graphics systems have only recently
become technologically and economically feasible, end
user chemists are now beginning to become computer
users. The following discussion is a case study of
the "chemists' computer revolution" at G. D. Searle.

Economics and practicality have been the driving force behind the
development of chemical documentation from its early days. The
"hieroglyphic"-type symbols in use in the mid-18th century could not
be accommodated by the printing press, which led Berzelius to
suggest in 1814 a nomenclature based on letters and numbers (see
Figure 1), where each element would be represented by the first
letter of its Latin name, and the number of atoms in the molecule
be designated by a number to the upper right of the elemental
symbol. When Berzelius later "simplified" his notation by using
bars, dots and commas to denote numbers of atoms, he only made it
worse. Liebig (1) proposed in 1834 that to facilitate printing,
superimposed notations not be used, and that the numbers be written
below and to the right of the element symbol (see Figure 2). The
familiar pictorial (valence) notation was first suggested by Couper,
and popularized by chemists of the day like Kekule (2).

Printing technology kept up with further development of the
two- and three-dimensional chemical graphs which not only aided
visualization of the molecules, but actually came to be depended
on by chemists to express their ideas. As the number of known
compounds grew, it became necessary to begin cataloging them.

0097-6156/87/0341-0018$06.00/0
© 1987 American Chemical Society

Verbal communication had necessitated the development of chemical
nomenclature, and its standardization made indexes of the chemical
literature possible. The earliest tracking systems depended on
conventional substructure methods with edge-punched cards.

The efforts in computer technology were based on non-graphic
business applications, so it was most practical for the chemical
information systems to be made to adapt to the standard business
computer. Systems evolved that were based on linear representations
of graphical formulas, such as Wiswesser Line Notation, WLN (see
Figure 3), or standard chemical nomenclature. But the linear
representations were one step further removed from the 2-D structure
that the chemist so depended on, and involved learning what in
essence was a foreign language. The chemists found themselves more
and more alienated from their own literature, and information
intermediaries found their place in the sun.

The Searle Case

That economics played a role early on at Searle was mentioned in a
paper my predecessor Dr. Howard Bonnett wrote in the Journal of
Chemical Documentation in 1962 (3). Work was being done at that
time on computer representation of 2-D chemical graphics, but it
required prohibitively expensive, single application, equipment. He
noted that one of the primary reasons Searle had adopted Wiswesser
Line Notation was that the notation could be used on a standard
accounting computer and did not require a high capital outlay.

When in the early Sixties Searle encoded its internal chemical
structures into Wiswesser Line Notation, chemists were for the first
time able to quickly discover the presence or absence of a compound
in the file, or what sort of analogues had been made of a certain
structure, first through alphabetic computer listings, then later
through permuted indexes and batch searching of the WLN. The
information scientist was an integral part of the process, encoding
the request, doing the search, and interpreting the WLN results into
the 2-D graphics the chemist could appreciate.

Through the years, many improvements were made to the original
system, until the late Seventies, when the "accounting" computers
were replaced by scientifically adaptable, interactive machines, and
our batch searching evolved to online retrieval using CRT's. The
WLN database itself had not changed significantly (other than to
grow), but it was augmented by the generation of WLN-fragments,
which allowed for a sophisticated online searching system (4). As
the system continued to be based on WLN, the information scientist
ran the searches, and the output continued to be interpreted for the
end user chemist.

The WLN-fragment search was notable in that its use did not
require full knowledge of the notation and its rules. A "reading"
knowledge of WLN was all that was required, and armed with a
WLN-fragment dictionary and a short training course, a few
adventurous chemists actually used the system themselves. It was
not for lack of enthusiasm that the system did not receive full

Ω Θ	hydrochloric acid	Scheele, *1772*
ΘX	calcium chloride	
 (P)U-	potassium sulfite	Lavoisier, *1787*
 (P)U	potassium sulfate	
 ⊕(L)⊕ ⊕	lead sulphate	Dalton, *1835*
O(Z)	zinc oxide	
KSO⁴	potassium sulfate	Berzelius, *1814*
K̇ S̈	potassium sulfate	Berzelius, *ca. 1830*
Ḧ	water	
Ḣ	hydrogen sulfide	

Figure 1. Early chemical hieroglyphs.

C₄H₁₀   butane   Liebig, *1834*

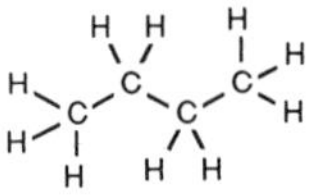

butane   Couper &

Kekule, *1858*

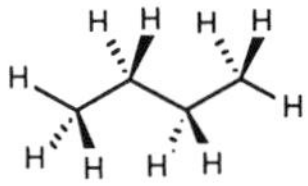

butane   contemporary

Figure 2.   Development of graphics-oriented representations.

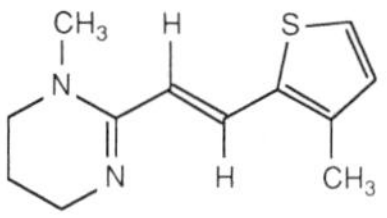

(E)-1,4,5,6-tetrahydro-1-methyl-2-

[2-(3-methyl-2-thienyl)ethenyl]

pyrimidine

T6N CN AUTJ C1 B1U1- BT5SJ C1

Figure 3. Linear representations of graphical formula:
standardized chemical nomenclature and Wiswesser Line Notation.

support from all the chemists. What prevented chemists at that time
from using the system was the nonavailability of computer terminals.
A big barrier yet to be overcome was the tendency of some in upper
management to view with suspicion any scientist's activity that did
not require beakers and flasks, and it was not until those basic
attitudes changed that the computer revolution for chemists could
possibly have happened.

The Move to Graphics

While end user chemists had never been involved in our in-house
chemical database development, when it became practical to look at
graphics alternatives, they were an integral part of the decision-
making. A task force of information specialists, systems analysts,
and chemists evaluated the course of action: whether to build a
graphics system in-house, or to buy existing software. Economic
factors continued to prevail, and it was seen that in-house
development, however much preferred from a customization standpoint,
would take too long to complete. The result of several months of
discussion, demonstrations and site visits, was the decision to
license Molecular Design Ltd.'s MACCS system, the now well-known
molecular connectivity-based retrieval system.

   Besides providing Searle with twenty years of access to and
organization of its compounds, the WLN turned out to be the crucial
link to the chemical graphics system towards which we had been
working. Ironically, it was at the time that we were replacing the
WLN that its strength became most evident. Though well-adapted to
computer manipulation, WLN was originally devised without this in
mind, as a rule-based, linearizing method of indexing structures.
WLN withstood the impassionate logic of the computer, and through
another purchased program, DARING, from Fraser Williams (Scientific
Systems) Ltd., our database of 50,000 WLN's was converted
automatically to connection tables with an error rate of less than
four per cent. Simple conversion of the DARING connection table to
MACCS connection table format and manual entry of the errors (about
2,000 compounds all told) allowed the entire database to be
converted and searchable in MACCS within seven months. At last, the
end user was able to use what has become the standard language of
chemistry to access internal chemical information. Information
scientists developed a full-day training program and users guide,
and trained 100 Searle chemists to use the new database by the end
of 1984, only nine months after acquisition of MACCS.

   That there was a change in attitude by R&D management toward
this first approach to end user chemist searching was due in no
small part to their investment in the decision. But success
required more than approval from above, more than a graphics
database, and more than a user-friendly system. What still needed
to happen was the proliferation of graphics terminals before end
user searching could become practical.

Hardware

Without convenient terminals and hard-copy devices, the situation
for the chemist would be little better than the early days of
chemical graphics, when even the largest of companies would only
invest in a few expensive workstations and chemists would have to
leave their laboratories to run a search. The real turning point
that made the computer revolution happen was that systems were
beginning to be hardware-independent and prices were continually
decreasing in the competitive computer market.

Simple graphics terminals like the Envision terminals that had
been used for database development had cost us around $8,000 each;
today there are comparable graphics terminals in the neighborhood of
$2,000. The standard VT-100 terminal just a few years ago cost
$2,000 but now can be had for about $600. To bring an entire
department into the computer age means a large capital investment
but the current pricing structure has allowed this to happen at a
much faster rate than could have been envisaged earlier.

The required terminal would have to be inexpensive, with
sufficiently high graphics resolution, would have to be compatible
with MACCS, but also be used for a variety of applications. Our
experience with the Envision terminal showed that it was not robust
or cheap enough for mass distribution. The Apple Macintosh was
chosen as the chemist's workstation since it fit the basic
requirements and had other advantages as well.

Choice of the Apple Macintosh

The small screen of the Macintosh was initially alarming, but after
a few minutes use, one realizes that the effective resolution is
sufficiently high to make the images clear and easy on the eyes.
The price of the Macintosh when we made our first purchases was
around $1600, less than the VT-100 a few years ago, and well within
the range of normal office equipment expenditures. Its compact
design makes the Macintosh fit well into cramped laboratory settings,
and the fact that it is portable aided its introduction to the
chemists, who were encouraged to take it home for practice.

The short learning curve is a distinguishing feature of the
Macintosh. Pull-down windows lessen the number of commands that
have to be remembered, and the use of a mouse rather than the
keyboard makes traditional typing skills less critical.

To make the Macintosh compatible with MACCS, we use the
Versaterm Pro software package, which allows the Macintosh to
emulate the Tektronix 4105, one of the acceptable terminal types.
By this same package, the Macintosh can emulate the Tektronix 4014
and VT-100, which allow the chemists to use other applications
programs, such as SYNLIB for chemical reaction literature and of
course, all the standard non-graphics applications such as
electronic mail.

The Macintosh can be hooked up to a high quality laserprinter (the Apple LaserWriter) via the AppleTalk network software, or when used as a terminal, it can direct output from systems such as MACCS or SYNLIB to a local laserprinter. (The Figures in this Chapter were done on the Macintosh and printed on the LaserWriter.) Another deciding factor in the choice of the Macintosh was the availability of quality software.

Software Examples

Once chemists became familiar with the Macintosh and started using the internal compound database, they began to branch out and discover their own computer applications. Chemical structures produced by MACCS, while fine for internal reports or correspondence, are not of publication quality. The ChemDraw software from Cambridge Scientific Computing, Inc., is now being used extensively for situations demanding high quality structures, such as slide presentations, and also for merging structures with text. Using a combination of ChemDraw and MacWrite, for example, the chemist can insert chemical structures into word processing documents. The Apple Switcher utility enables the chemist automatically to switch back and forth among several programs to create the desired report. Sample output from this simple process is shown in Figure 4. The text was written using MacWrite, with the full Macintosh complement of fonts, styles, special features like bolding, etc. The chemical reaction sequence was drawn using ChemDraw.

Some of the chemists' demands for structures are based on specialties, for instance, peptides, where the need is for a hybrid notation rich in text but with some structural elements. Figure 5, for example, shows the well-known neuropeptide vasopressin, drawn with ChemDraw.

For chemical reaction literature, chemists use the SYNLIB database from Distributed Chemical Graphics. SYNLIB is a well-documented, user-friendly system, designed for end user browsing. Figure 6 shows sample SYNLIB printed output, two records to the page. For chemical supplier information, the Fine Chemicals Directory from Fraser Williams (Scientific Systems) Ltd. is available through MACCS, as is an internally developed database of chemicals available in our Chemical Stockroom.

Molecular Modeling

For those with an interest in true 3-D structures, we have also a number of user-friendly molecular modeling packages available on our system, among them SYBYL, from Tripos Associates, Inc., and Macromodel, from Prof. Clark Still, Columbia University. Both can be accessed via the Versaterm Pro emulation software on the Macintosh, as well as from intermediate and high-performance workstations like the NEC-APC and the Evans and Sutherland PS-300. These specialist packages are maintained by our Drug Design department, who help out users with the specifics of the software.

An abstract in a recent **CA Selects** drew my attention to the fact that enzymes can be used to hydrolyse hydantoins to amino acids under mild conditions, and in many cases can selectively convert DL starting materials to pure D or L products, often quantitatively:

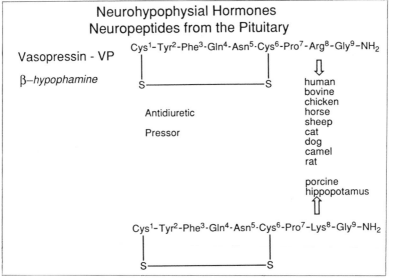

When one of the substituents is hydrogen some of these enzymes function as both aminohydralases and as racemases, leading to 100% conversion of racemic hydantoin to just one amino acid enantiomer; others selectively hydrolyse one hydantoin enantiomer leaving the other unchanged. Whether the pure D or L amino acid is produced depends on the particular enzyme system involved. The second step, cleavage of the intermediate N-carbamoyl aminoacid to the free aminoacid can be enzymatic or chemical, but in either case is achieved without racemization under relatively mild conditions.

I had Information Services do a CA search on this topic and a copy of the 21 references found is attached. From the number of recent patents on the subject it would appear that this has become a method of some industrial importance, especially for the production of optically pure unnatural and D-amino acids.

The actual experimental conditions employed range from the use of cultured broths of common microorganisms, through the use of cell-free extracts, to the use of columns of fully immobilized enzymes embedded on cellulose triacetate fibres, and reactions are rapid at 30°.

Figure 4. Example using the Apple Switcher Utility: text done in MacWrite, reaction sequence in ChemDraw.

Neurohypophysial Hormones
Neuropeptides from the Pituitary

Vasopressin - VP

β–hypophamine

Cys^1-Tyr^2-Phe^3-Gln^4-Asn^5-Cys^6-Pro^7-Arg^8-Gly^9–NH$_2$

S————————————S

⇩

Antidiuretic

Pressor

human
bovine
chicken
horse
sheep
cat
dog
camel
rat

porcine
hippopotamus

⇧

Cys^1-Tyr^2-Phe^3-Gln^4-Asn^5-Cys^6-Pro^7-Lys^8-Gly^9–NH$_2$

S————————————S

Figure 5. Example showing the versatility of ChemDraw.

SYNLIB ™ V2.2   27-AUG-86 16:42

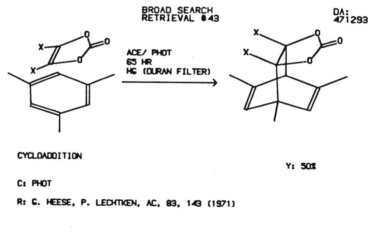

BROAD SEARCH
RETRIEVAL #33   DA:
                                                          398797

A. (I)/ ET₂O/ ZN
   ADD (II)
B. 2 HR/ RT

R = ET

Y: 80%

C: OMET

R: I I LAPKIN, R M KISLOVETS, ZOK, 4, 801 (1968);
   CA, 69, 19054 (1968)

BROAD SEARCH
RETRIEVAL #43   DA:
                                                          471293

ACE/ PHOT
65 HR
HG (DURAN FILTER)

CYCLOADDITION

Y: 50%

C: PHOT

R: G. HEESE, P. LECHTKEN, AC, 83, 143 (1971)

Figure 6.   Sample output from SYNLIB.

Conclusion

Bringing the chemists into the picture by bringing pictures to the chemists' laboratories has brought about a revolutionary change in the way the chemists do their work. It has happened at Searle through a combination of events, in some cases quite fortuitously. It depended on a needed 2-D structure database and user-friendly system, on the approval of R&D management for end user searching, on cheap, multi-purpose graphics terminals, on user involvement and training, and on the chemists' cultivating the habit of using a computer terminal.

     This last point is very important. People who work with computers day in and day out are in the habit of logging on, reading their mail, and doing their work. Scientists who use the computer only as an adjunct to their work need opportunity to stay familiar with it, such as regular use of electronic mail, or keeping private files up-to-date. For such casual users, the retraining curve must not be too steep, or they will be discouraged. Of course, this is where an online HELP system or pull-down windows can prove invaluable.

     The part of the information scientist in the revolution should not be underestimated. For guaranteed, all-around success, someone had to work out all the details beforehand, from establishing a quality database, to selecting the hardware and initial software, to providing indepth training. Information scientists are in the unique position of having their heart in the subject matter, as well as knowing the computer systems. There has been a change in their role as well. They are not in the middle of every internal search request, but the more complex questions continue to be referred to them. Information scientists continue to do the database maintenance and development, while taking on additional responsibilities for training and "customer support", and evaluation of new systems and applications.

     While a quantum leap has been made, there is still much to be done. Medicinal chemists want to see structures and data together, and while we automatically can get biology data from results of a substructure search, or get structures printed for compounds with certain biology data parameters, our current system is not flexible enough and the databases are not truly integrated.

     While the chemists have tools, like ChemDraw, to produce chemical structures to their specifications, they still have to do it themselves. Since beauty is in the eye of the beholder, there will always be an artistic (subjective) dimension to chemical structure drawing that will make complete automation inconceivable. For example, the Merck Index shows the benzomorphan Metazocine as a highly stylized depiction of this bridged tricyclic structure, which represents only one of many opinions as to how it should be drawn (see Figure 7).

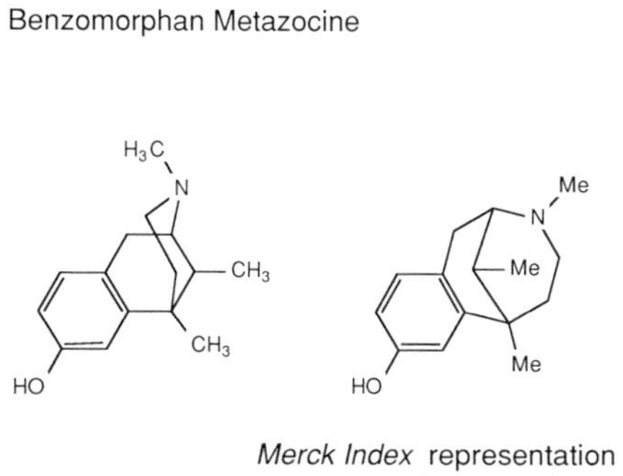
Benzomorphan Metazocine
H3C
N
CH3
CH3
HO
Me
N
Me
Me
HO
Merck Index representation

Figure 7. Two graphical representations of the same compound.

Another deficiency is that a single software package may not satisfy all requirements. Structures stored in MACCS may not be drawn in a preferred way, so a package like ChemDraw is used. The Switcher program was written because several packages may need to be used to get the desired result. The chemists' requirements are severe, and we have not met them all.

The theme of this Chapter is bringing computer graphics to the laboratory. Perhaps the natural consequence of this will really be to reunite chemists with their literature. While it may not be economically viable now for end user searching of commercial databases, there are alternatives, such as optical disks, that are being developed for low-cost end user browsing, which in the future might provide in-house access without the risk of cost overruns. By helping chemists become computer literate, we are approaching that goal.

Acknowledgments

The author would like to thank Michael Clare for sharing his expertise, for contributing helpful suggestions, and for providing Figures 4 and 5.

Literature Cited

1. Jorpes, J. E., _Jac. Berzelius, His Life and Work_; Alqvist & Wiksell: Stockholm, 1966.
2. Pauling, L., _General Chemistry: An Introduction to Descriptive Chemistry and Modern Chemical Theory_; W. H. Freeman: San Francisco and London, 1953.
3. Bonnett, H. T.; Calhoun, D. W., _J. Chem. Doc._ 1962, _2_, 2-6.
4. Johns, T. M.; Clare, M., _J. Chem. Inf. Comput. Sci._ 1982, _22_, 109-113.

RECEIVED March 11, 1986

Chapter 4

Microcomputer-Based Software for Chemical Structure Management: A Comparison

Daniel E. Meyer

Chemical Information Division, Institute for Scientific Information, 3501 Market Street, Philadelphia, PA 19104

Four microcomputer-based structure management software
programs were compared for hardware requirements and
system features. These programs, ChemBase, ChemFile,
ChemSmart, and PSIDOM, offer the capability not only
to draw and store chemical structures, but also retrieve
records via structure, substructure and text search
techniques. The graphic quality of the records, system
expertise, and price vary considerably among these
products and as such offer the user a wide selection
of products to choose from to meet his needs.

In the past few years, several microcomputer-based software packages
have become available to input and store chemical structures. These
packages can be divided into two main groups: those that allow structures to be drawn and stored for later recall [e.g., ChemDraw (S.
Rubenstein), WIMP (Aldrich)] and those that capture structures from
online files like Index Chemicus Online and CAS ONLINE [e.g, PC-Plot
(MicroPlot), Emu-Tek (FTG Data Systems)]. The greatest limitation
of both groups of software is that stored structures have to be recalled serially or by specific registration number since structure
and sub-structure search capabilities are not available.

In 1986, several microcomputer-based software packages have
become available which allow not only the creation of chemical
structure files, but also the capability to search for specific
structures, substructures and text from the files. Four of these
newly available packages, ChemBase, ChemFile, ChemSmart, and PSIDOM,
were reviewed and the basic system features compared.

A fifth package in this structure management category, HTSS
(LexTex, Inc.), has been recently announced but has not been evaluated. HTSS will be included in later reviews.

Software Descriptions

ChemBase. ChemBase, from Molecular Design Limited, is a very powerful structure and reaction management program which is part of a
series of programs called the Chemist's Personal Software Series

(CPSS). As part of this group, ChemBase interfaces with ChemText, a text editor program which allows for graphics to be easily inserted into text and ChemTalk which is a communication/graphic emulation program to interface with Molecular Design Limited's mainframe software systems. As can be seen in Table I, ChemBase requires a minimum of 640K memory and a mouse input device. An initial database of structures and reactions is provided with the software as well as a large library of structure templates. Chem- Base provides considerable system expertise by checking for proper valence, automatically generating molecular formula and molecular weight, and highlighting atom overlap. These features facilitate the input of structure records or queries. A sample structure display from ChemBase is shown in Figure 1. (All figures were produced on a dot matrix printer rather than a laser printer to show the output quality from these programs on standard office/ laboratory equipment.)

ChemFile. ChemFile, from COMPress, is a very easy to understand structure management package which is marketed with a similar pack- age called ChemLit. The main difference between these two packages is that ChemFile allows for text to be entered in specific fields while ChemLit has a single note field where free-text notes and abstracts can be entered and then retrieved via string-searching. A feature of ChemFile's field format is the ability to search for ranges of numeric data such as molecular weight, boiling point, etc. However, each data field is restricted in size to less than 20 char- acters. ChemFile works easily on a two-floppy, 256K system and when searching a file, asks if the user want to continue the search on another diskette. This feature is a great help in allowing user files to grow beyond the storage limits of a single data diskette. The software comes with two versions of the program: one for using the cursor keys and the other version for using a mouse. ChemFile has an interesting input feature in providing a string interpreter which allows the entry of common line formulas of standard substi- tuent groups, e.g., COOH is translated into -$\overset{\text{O}}{\underset{\text{}}{\text{C}}}$-OH. A sample display from ChemFile is shown in Figure 2.

ChemSmart. ChemSmart, from ISI Software, is an easy-to-use pro- gram which is the first product from a series of microcomputer-based products designed to handle chemical data. ChemSmart allows for structure/reaction input and has a corresponding note card for each entry to store data such as compound name, molecular formula, molec- ular weight, physical/chemical properties, etc. Of the text, only the molecular formula, name and registry number fields are search- able. Different styles of notebooks are provided for structure and reaction data. ChemSmart works easily on a two-floppy, 256K system and the single program diskette handles either cursor or mouse controls. An interesting feature is the "Bond Compass", a direct- ional guide which assists in the rapid input of structure forms. A database of 250 structures and a sample reaction is provided with the software as well as a number of standard templates. ChemSmart provides limited system expertise by checking for proper valence when structures are drawn. A sample structure display is shown in Figure 3.

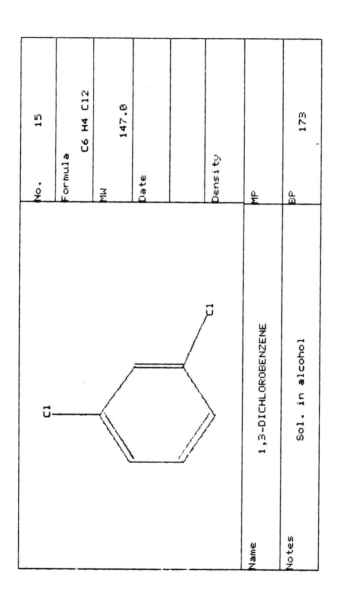

Figure 1. Sample Display from ChemBase.

Table I.   Hardware Requirements

	MACHINE TYPE	MEMORY	DRIVES	MOUSE	GRAPHICS BOARD
CHEMBASE	IBM	640K	TWO FLOPPY DISK DRIVES (360 KB EACH) (HARD DISK DRIVE RECOMMENDED)	REQUIRED	IBM & COMPATIBLE, HERCULES
CHEMFILE	IBM	256K	TWO FLOPPY DISK DRIVES (HARD DISK OPTIONAL)	OPTIONAL	IBM & COMPATIBLE
CHEMSMART	IBM	256K	TWO FLOPPY DISK DRIVES (HARD DISK OPTIONAL)	OPTIONAL	IBM & COMPATIBLE
PSIDOM	IBM	512K	HARD DISK DRIVE	OPTIONAL	IBM & COMPATIBLE, HERCULES

```
name 1,3-DICHLOROBENZENE
CN 15
MF C6H4Cl2
MW 147.01
```

```
Hit 1 of 1.   Continue, y/n?
```

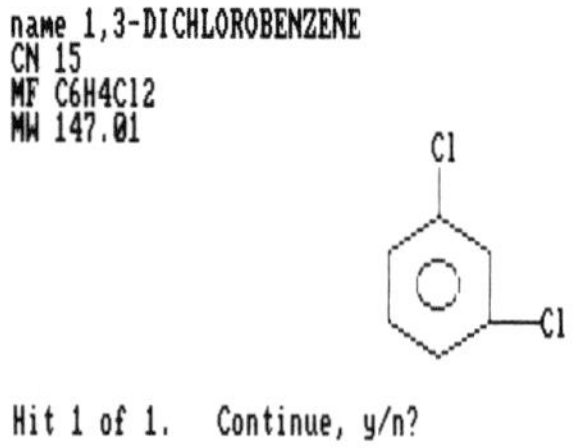

Figure 2.   Sample Display from ChemFile.

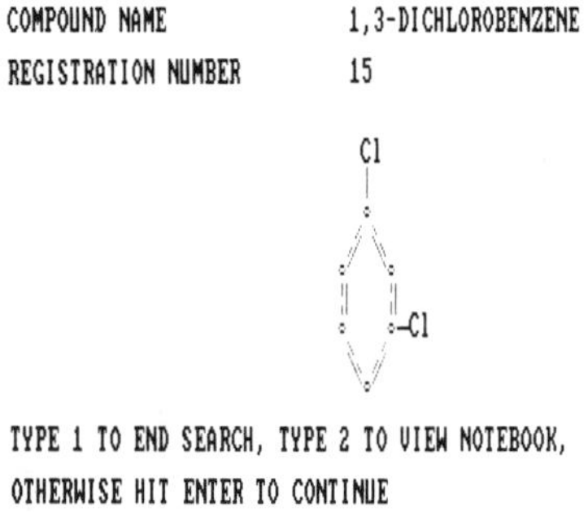

```
COMPOUND NAME          1,3-DICHLOROBENZENE
REGISTRATION NUMBER    15
```

```
TYPE 1 TO END SEARCH, TYPE 2 TO VIEW NOTEBOOK,
OTHERWISE HIT ENTER TO CONTINUE
```

Figure 3.   Sample Display from ChemSmart.

PSIDOM. PSIDOM, from Hampden Data Services, is a series of modules
for handling chemical structure and text data. The basic module
for structure input and display is called PsiGen. A second module,
PsiBase, is now available which permits structure/substructure
searching against PSIDOM-created files. Both of these programs
will be treated as one unit so as to be directly comparable with
other packages reviewed. Other PSIDOM modules are available to
integrate text files with structures (PsiText), print out connection tables (PsiRep), and so forth. PSIDOM requires a hard disk and
a minimum of 512K memory. A mouse or cursor controls can be used
for input. PSIDOM has a broad range of input techniques ranging
from standard free-hand drawing to convenient ring and chain
command notations. Feldmann notation input (e.g., 66U6D5 for a
steroid skeleton) is also available which allows for the rapid
building of large ring systems. PSIDOM also allows for substituent
group shortcuts as discussed with ChemFile. The system automatically generates molecular formulas and has limited valence checking.
A sample structure display is shown in Figure 4.

System Descriptions

Hardware Requirements. Table I lists the basic hardware requirements for the four programs. ChemBase and PSIDOM require a hard
disk drive and a minimum of 512K memory (640K is recommended for
ChemBase). ChemFile and ChemSmart work well with a standard two-
floppy machine with 256K memory. Only ChemBase requires a mouse
input device but all four programs can utilize the mouse.

Data Input Capabilities. All of the programs permit the standard
input features of different atom and bond types. Table II compares
a few of the input features and shows that all but ChemFile provide
the additional stereochemical bond representations. All of the
programs allow for templates to be used for input and all but
ChemFile provide templates with the package. Input methods differ
greatly as ChemBase offers standard modes plus rubberband effect,
ChemSmart uses a directional Bond Compass, and PSIDOM uses a range
of shortcuts and notation commands.

Manipulating Structures. Table III shows that all of the programs
offer the standard features necessary to arrange structures on the
screen. ChemBase, however, is the only program which offers a very
powerful "Clean" command which takes a poorly drawn structure and
generates a "clean", chemically correct version.

Searching Capabilities. The searching capabilities, which make
these programs different from previously available drawing programs,
are shown in Table IV. ChemBase has the most powerful search capability which includes all text fields and the ability to retrieve
a structure from a reaction scheme specifically as either a reactant
or product. ChemFile has a range feature for each specific numeric
field but is limited in the amount of data that can be stored (less
than 20 characters per field).

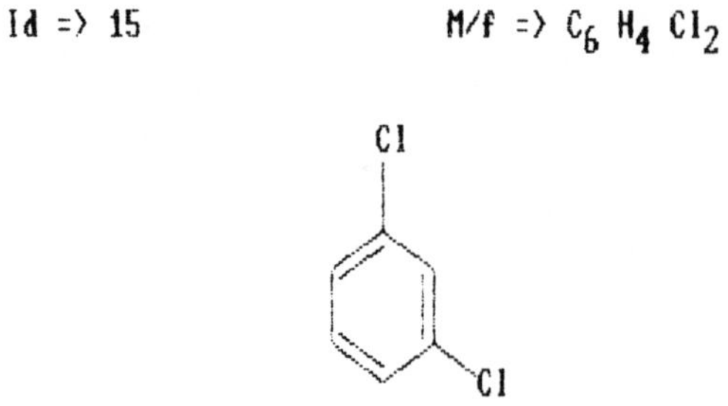

Id => 15   M/f => $C_6 H_4 Cl_2$

Figure 4.   Sample Display from PSIDOM.

Table II.   Data Input Capabilities.

	TEMPLATES	DRAW MODE	STEREOCHEMISTRY
CHEMBASE	Y	NORMAL CONTINUOUS RUBBER BAND	UP DOWN EITHER
CHEMFILE	NONE PROVIDED (FROM USER STORED STRUCTURES)	CONTINUOUS RUBBER BAND	(DOTTED LINE ONLY)
CHEMSMART	Y	BOND COMPASS	UP DOWN
PSIDOM	Y	CONTINUOUS RING AND CHAIN COMMANDS FELDMANN NOTATION COMMANDS	UP DOWN EITHER

Data Provided with Software. Table V shows the amount and types of
data provided with each product. ChemBase and ChemSmart both
provide sufficient databases and templates to give initial search
examples and increased utility to the user. PSIDOM provides a
large supply of templates and all of the programs allow stored
structures to be recalled as templates.

Prices. The price of these packages is an equally important factor
in deciding which package can suit your needs, but the prices listed
here can only be used as a relative guide. Due to price changes,
upgrades in software, different pricing policies for international
territories, and academic discounts, it is quite possible that some
of the prices listed below are not the current price. However, I
did not want to eliminate this information from the paper since it

Table III. Manipulating Structures

	ENLARGE	SHRINK	CLEAN	DELETE	ROTATE	MOVE
CHEMBASE	Y	Y	Y	Y	Y	Y
CHEMFILE	Y	Y	N	Y	N	Y
CHEMSMART	Y	Y	N	Y	Y	Y
PSIDOM	Y	Y	N	Y	Y	Y

Table IV. Searching Capabilities

	EXACT STRUCTURE	SUBSTRUCTURE	TEXT	REACTIONS
CHEMBASE	Y	Y	Y	Y
CHEMFILE	Y	Y	Y	N (AVAILABLE ON CHEMLIT)
CHEMSMART	Y	Y	LIMITED TO MF AND NAME	Y
PSIDOM	Y	Y	LIMITED TO NAME	REACTION SCHEME DISPLAYABLE ONLY

Table V. Data Provided with Software

	STRUCTURES	TEMPLATES	REACTIONS
CHEMBASE	100	57	100
CHEMFILE	NONE	NONE	NONE
CHEMSMART	250	9	1
PSIDOM	NONE	80	NONE

can be a factor in the decision-making process. The current prices
(as of January, 1987) for the four software packages are as follows:

ChemBase $3,500 ($975 academic price)
ChemFile $ 150
ChemSmart $ 335
PSIDOM $ 600-1600 (depending on supporting
                            modules purchased)

Summary

All of these packages offer the capability to build personal databases and then recall structures via registration number, text, and
structure. This last feature, structure search capability, makes
these packages different and more powerful than the structure drawing or data capturing programs which have been available for the
past few years. Some of the existing drawing programs will most
likely be modified to incorporate search capability in the near
future. PSIDOM was first released as a personal structure file
manager with individual record recall and now has been expanded
to permit structure and substructure retrieval against its stored
records. More entries into this market seem likely in the next
12-18 months.
   The advent of microcomputer-based systems such as those described here provides a new opportunity for individuals to store and
manipulate personal and proprietary data. Small data collections
of up to a few thousand compounds can be stored to keep track of
new research, inventories, and laboratory data. Such information
has traditionally been kept in card files or unindexed file folders.
When stored as a PC-based structure file, this data can be
rapidly searched and specific information easily retrieved and
evaluated.
   Several papers have appeared recently which describe how
personal structure management systems can be used for structure
retrieval (1) or utilized to search for structure-activity relationships (2). The lower cost of the microcomputer environment makes this
technology affordable to a larger audience, including individual
researchers and students. These four programs, ChemBase, ChemFile,
ChemSmart, and PSIDOM are quite different from each other in their
graphic quality, system expertise, and price and as such offer the
user a spectrum of products to review before selecting the one which
meets his needs.

Literature Cited

1. Contreras, M. L.; Deliz, M.; Galaz, A.; Rozas, R.; Sepulveda, N.
    J. Chem. Inf. Comput. Sci 1986, 26, 105-108.
2. Meyer, D.; Cohan, P. Am. Clin. Prod. Rev. 1986, 5(12), 16-19.

RECEIVED March 17, 1986

Chapter 5

An Electronic Notebook for Chemists

John Figueras

65 Steele Road, Victor, NY 14564

Chemical information systems are now available for building personal
databases on personal computers. They may be used as stand-alone,
personal information systems, or as front ends for accessing
commercial databases. The programs offer facilities for handling
molecular diagrams with accompanying data or text. Powerful search
capabilities offer retrieval based on data, chemical substructure or full
structure. Interactive graphics for structure entry make these
programs easy to use.

The advent of microcomputers with ample memories, fast disk drives, convenient
graphics input devices and reasonable processor speeds has permitted development
of structure-based chemical data systems that were formerly restricted to large
mainframe systems. A factor that has been very important in this development is
the availability of modern, well-structured languages such as C and Pascal, which
can be compiled to native machine code for high execution speeds. A particularly
apt language for programming a chemical-structure-based systems is Turbo Pascal,
which provides a well-structured language (a necessity for writing complex
programs) and a number of low level routines that give direct access to the video
screen for high speed, flexible graphics manipulations. The price paid for using
these low level routines is that the programs that I will be describing will run only
on PC's that are 100% IBM PC compatible.

A number of PC-based structure-handling programs with various capabilities
have appeared recently in the literature (1,2) and as commercial items. ChemSmart,
a system sponsored by Academic Press, offers a chemical information file
management system with small databases from the Institute for Scientific
Information. ChemBase offered by Molecular Design Ltd (MDL) can operate as a
stand-alone system, or can interact with databases created by MDL systems on a
mainframe computer. ChemFile and chemLit, subjects of this article, are trade
names of two suites of programs available from COMPress (a division of
Wadsworth Publishing). All of the programs mentioned above have many features
in common, but differ in details of price, access to commercial databases, quality of
graphics, flexibility and ease of use.

0097-6156/87/0341-0037$06.00/0

ChemFile and chemLit were written for an IBM PC having 256K or more of memory. As implied by the title of this paper, the programs were written to keep track of chemical data, with emphasis on recording and retrieval of chemical structure diagrams. ChemFile and chemLit share common structure entry front ends and common substructure search capabilities, but differ in data-handling capabilities: chemLit is oriented towards text data, and chemFile is oriented towards numerical data. These differences in data types give rise to differences in information retrieval capabilities, which will be described later. At this point, neither program has facilities to interface with commercial databases.

The program suites are distributed in two versions: one uses a mouse for graphics structure input, and the other uses the keyboard cursor control keys. These devices drive a cursor on the screen for selecting menu items and placing pieces of structure on the screen. A mouse, now available through mail order houses for about $100, is strongly recommended for ease of use. Aside from the input device, the two versions are identical in every respect.

With these programs, the chemist can draw a structure on the PC screen and attach chemical information to the structure. Depending upon the program (chemFile or chemLit), 'information' may be literature references, chemical stockroom inventories, laboratory results, biochemical data, results of product evaluations, and so on. ChemLit has the further capability to handle chemical reactions. The data, with structure, may be stored on floppy disks or on a hard disk. A database may extend over a series of floppy disks, the maximum number being nine hundred and ninety-nine, and the series can be searched sequentially from a given query. Hard disk storage can be as much as 36 megabytes. A single floppy disk can hold 300-400 structures, depending upon the amount of space used for data and the size distribution of the structures in the database.

Structure Input

Figure 1 shows a structure created in chemFile and the main structure entry menu. In developing the structure entry program, two design *desiderata* were imposed:

1. The structure entry menu must be simply constructed and simple to use.

2. The program must produce good-looking structure diagrams.

Accordingly, the structure entry menu is restricted to one line at the top of the screen, as shown in Figure 1, and all sub-menus are similarly constructed. The second line of the screen is reserved for one-line, terse prompts that instruct the user in what to do next. These prompts can be ignored by the experienced user. The prompts, which change as each operation is completed, provide subliminal feedback to let the user know that the current operation has been carried out. In addition, context-sensitive help screens are available for all menu or sub-menu items.

Much effort went into design for good looking structures. In particular, nearly-vertical and nearly-horizontal lines are automatically justified to their correct positions to eliminate the jagged lines that occur on a raster-based display. Care was taken that atoms were properly centered at the ends of bonds, and that pendant ring assemblies such as diphenyl have rings properly lined up. A special character set was constructed to create subscripts, and to generate the pi-circle in a benzenoid ring.

The structure-building menu items used for structure entry are reviewed in brief, providing a scaffold to demonstrate the structure representation capabilities of the programs. A menu item is selected by rolling the mouse to move the cursor to the desired item; the left button is clicked to make the selection. If the righthand button is clicked, a help screen pertinent to that menu item appears. The help screen can be accessed at any time, even with a structure in the work area of the screen. In the following description of the structure-building items, refer to Figure 1.

The DRAW item is used for free-hand drawing. Drawing can be done so that all bonds automatically will be of uniform length, or this restriction can be removed by toggling the FREE option at the left side of the menu. DRAW is used to create stick diagrams, and may be used to join pieces of structure. A rubberband line helps the user visualize the drawn line.

Rings are drawn without the use of templates. The chemist selects a ring size from the RING sub-menu and uses mouse accesses to select the location of the first edge of the ring. The ring is automatically completed in a clockwise direction by the program. The orientation of the first edge determines the orientation of the ring. If the first edge is already in a ring, a fused system is obtained. Rings appear with fixed, uniform edge lengths. The benzenoid ring option produces a six-membered ring with a centered pi circle. The program recognizes this ring as aromatic; it does not recognize Kekule` forms as aromatic. Aromatic rings are represented internally as six-membered rings whose elements are connected by an aromatic bond type. Later versions of the programs will extend the application of aromaticity to rings of different sizes. The aromatic bond type is assigned by the program when the benzenoid option is selected from the RING sub-menu. It is available as a separate bond type in the substructure search menu to allow specification of "don't care" connections of aromatic type.

A troublesome problem in putting structures together is getting pendant rings to line up properly, as in diphenyl. This is handled as follows. The DRAW option is used to create a horizontal or vertical bond from one of the rings. A new ring is started from the 'free' end (actually, there is a carbon atom there) of this bond. The second point required to define the first edge of the new ring is located approximately along the correct angle for that edge. The program senses this situation and computes the correct angle required so that the new ring will line up properly (horizontally or vertically) with the previous ring. An example of the result is shown in Figure 2.

A single bond is the basic bond created by DRAW, RING (and the item GROUP not yet considered). The BOND option allows the user to select a bond type from the BOND sub-menu and change the nature of any explicit, non-aromatic bond in the structure to the selected type. The endpoints of the bond to be altered are accessed by the mouse after a bond type has been selected. A dotted bond type can be used for limited description of stereo arrangement. A bond-erase feature is built into the sub-menu and is obtained by selecting bond-type zero. Examples of the different bond-types appear in Figure 1.

One of the most flexible input options is the GROUP option. This activates the keyboard for input of chemical strings. The program 'recognizes' almost any legitimate string of chemical symbols. This is made possible by a string interpreter that converts the string into a meaningful internal representation of structure. For example, a sulfonyl group input as a string 'SO_2' and placed in a structure, is processed through the interpreter, which generates a representation of this group as a central sulfur atom with two doubly-bonded oxygen atoms. Moreover, any groups attached to the SO_2 group, as in $CH_3SO_2CH_3$, are properly distributed with respect to the sulfur atom. Interpretation of strings is necessary to allow compatibility among numerous representations of functional groups, which is particularly important in structure-based searches.

GROUP serves a number of different purposes. It is used to replace atoms in a structure, facilitating representation of heterocycles. Stand-alone strings can be placed on the screen, presenting convenient input of structures easily represented by a linear string, such as the dimethylsulfone example considered in the previous paragraph. GROUP offers an atom delete function, implemented by a single space as a replacing 'atom.' A fast and easy path for building complex structures can be based on GROUP alone. Figure 3 shows a structure built up in this fashion solely from strings. The appearance in this example of a hyphen for a single bond and an equal sign for a double bond displays another facet of the interpreter, which properly interprets these symbols as bonds of proper type.

The GET option retrieves a structure in the database from disk, using a name previously assigned. The returned structure can be modified and stored with its own data set. This is a useful way to capitalize on previous work for building a complex structure.

STORE transfers a newly built structure to the database on disk. When STORE is invoked, the program first moves into a data entry mode. At this point, the user is asked to provide a name for the structure, which is required for subsequent GETs and for editing procedures described later. The data entry phase that follows is somewhat different for chemFile and chemLit, and is described in the next section.

FORMAT offers geometric transforms for translation and size rescaling. These are used to keep the structure properly placed within work area boundaries. The PLOT option directs structure output to a high resolution dot matrix plotting routine for printer output, or, optionally, to a pen plotter for high-quality drawings.

Data Entry, chemLit

The data entry opportunity for chemLit occurs when the STORE option is requested in structure entry. ChemLit data are completely text-oriented. A blank area in the work space below the structure is used for text input. Up to 13 lines of input, 79 characters per line, are available, depending upon the amount of space consumed by the structure. How this text space is used is entirely a matter of user choice. It may be used for brief abstracts, descriptions of chemical preparations or reactions, examination questions, and so on. After text is entered, the record is stored in the database, and becomes available for later searches.

ChemLit has facilities for input of chemical reactions. A reaction is indicated by a reaction arrow ,the composite '->', placed in the GROUP option between two structures. The user may mark the atoms that actually participate in reaction. Later searches can be made for the flagged atoms to retrieve specific reactions.

Data Entry, chemFile

Data entry in chemFile takes place in the STORE operation. The example in Figure 4 shows an almost completed record ready to be stored. Field labels identify the input lines with field lengths indicated by underscoring. Definitions of field labels, field lengths and data types are set up by the user in a utility provided by chemFile

free
[OFF]
draw
ring
bond
group
store
get
format
plot
QUIT
O
CH2COCH3
O

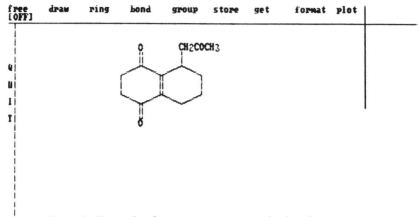

Figure 1. Example of structure entry menu (top) and structure.

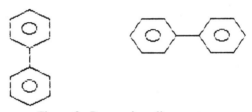

Figure 2. Proper ring alignment.

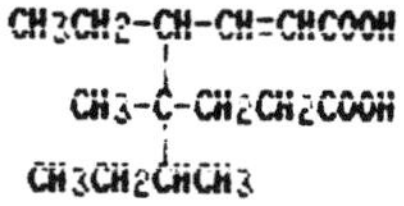

CH3CH2-CH-CH=CHCOOH
CH3-C-CH2CH2COOH
CH3CH2CHCH3

Figure 3. Structure building based on strings.

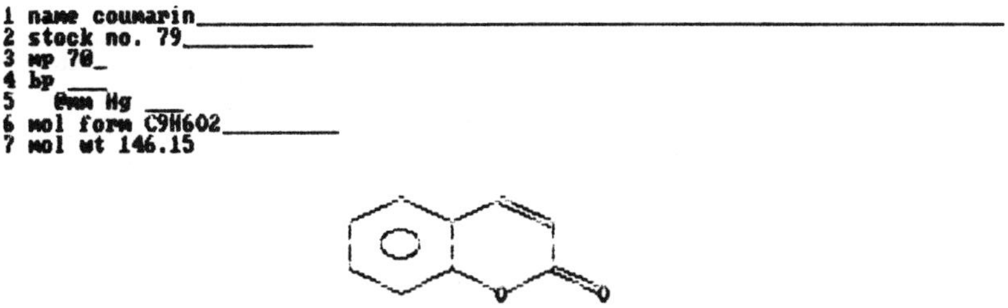

1 name coumarin
2 stock no. 79
3 mp 70
4 bp
5 @mm Hg
6 mol form C9H6O2
7 mol wt 146.15
O
O

Figure 4. Data Entry in chemFile.

that initializes the database. Initially, the nature of the data fields in chemFile is completely open: the user may define data fields to fit chemFile to a particular application. As many as 20 fields can be set up with integer, decimal or nonnumeric data types. ChemFile is primarily number-oriented so that values can be compared for magnitude in data searches. It has limited capability for handling characters. For example, one could set up a color field that would accept standard color names. Reaction indexing, which is more likely to require the descriptive text capability inherent in chemLit, is not available in chemFile.

Substructure Searches

Substructure searches are implemented in 'standard' fashion: that is, a fast screen-out is used to get rid of obvious nonmatches, followed by a full atom-by-atom search on the structures that pass the screen. The atom-by-atom search is a backtracking, graph-tracing procedure, resembling those used by Chemical Abstracts Service and Molecular Design Ltd.

Substructure search queries are entered from a substructure entry menu similar to the structure entry menu described above. The substructure entry menu differs in the absence of disk-storage routines, geometric transforms and plotting routines, and its BOND option is extended to allow placement of partial (open-ended) bonds to indicate "don't care" connections, and specification of bonds as aromatic. All valences at atoms are assumed filled unless otherwise indicated by the attachment of an partial bond (What You See Is What You Get). For this reason, special atom symbols are provided as 'generic' atoms. The symbol 'Z' is generic carbon, <u>any</u> carbon atom regardless of substitution. Similarly, the symbol 'X' is generic noncarbon. The introduction of generic atoms allows more general specification of substructures for searches; for example, the search query ZCOZ would specify a search for all ketones, regardless of substitution on the alpha carbon atoms. Searches for the simultaneous occurrence of a number of substructures are established by placing all of the substructures simultaneously on the substructure search screen.

The chemLit substructure search menu permits flagging of atoms that participate in reactions. One could, for example, retrieve all instances of conversion of esters to primary alcohols. Searches can be done not only for specific reactions, but for such queries as "Give me all reactions in which an aldehyde is a product" or "Give me all reactions that use aromatic nitro compounds as starting materials."

Data Searches

A search for chemFile data is submitted on a form created by the program, using the field attributes declared by the user when the database is first set up. As exemplified in Figure 5, the form provides for searches over <u>ranges</u> of data. If a specific value is required, a single value is entered into either one of the range fields. The request shown here sets up a search for non-toxic solvents boiling in the range 100-150 degrees.

Text searches in chemLit can be based on whole strings or on substrings, and can be carried out on the title of the record, or on the text body of the record. The searches are case-insensitive. Because chemLit's data are composed of strings, searches over numerical ranges (as in chemFile) are not possible. The search logic is very flexible: input consists of groups of items that are AND'ed together, and the AND groups are OR'ed together. Any number of items may comprise an AND group, and any number of AND groups may be OR'ed. Negative searches are

```
Specify Search Values at the Prompting "?".  Prefix a "*" for substring searches.
Use substrings only with IDs!  Use multiple substrings in NAME field only.
name -----------------------------------------------------------------------------
stock no. ---------------

                  mp lo limit ---        hi limit ---
                  bp lo limit 100        hi limit?150
            @mm Hg lo limit ---          hi limit ---
          mol form lo limit ---------    hi limit ---------
            mol wt lo limit -------      hi limit -------
        appearance lo limit -----------  hi limit -----------
 toxicity, h, m ,l lo limit -           hi limit -


***To skip any field, press ENTER alone
```

Figure 5. Data search specification in chemFile.

possible by placing a minus sign in front of the NOT item. Adding to the flexibility of this text search is the fact that substrings may be used. For example, a search on the word 'toxic' brings back all records containing variations such as 'toxicity' and 'nontoxic.'

Other Search Capabilities

Figure 6 displays the main search menu, and is introduced at this point because it contains other search features not yet discussed. One of these features represented in items 3 and 4 on this menu is the ability to do cascaded searches. Each search, whether it is a substructure search, a data search (chemFile) or a text search (chemLit), stands as an independent entity. There are occasions, however, when one would like to have interdependent searches. For example, one might like to refine a substructure search using the results of a previous search rather than the whole database. This capability might be important with large databases. Or, one might like to cross a data search with a substructure search, effecting a logical AND between the searches. An example of the latter would be a search for all nitro compounds having a musk odor, based first on a substructure search for nitro compounds followed by text search for the word 'musk.' Such searches are implemented using the record of hits from the last search as the basis for the next search. This gives a new record containing the mutual hits from the last two searches. Any number of cascaded searches may be carried out in this fashion.

The user may want to look at results from a previous search in several different ways, that is, to conduct several cascaded searches on the same hit record. A cascaded search changes the hit record; therefore, the program allows the user to retain a back-up copy of a given hit record by means of the SAVE option on the menu. The base record can be restored for a new cascaded search with the option RESTORE. The SAVE record is maintained until the next SAVE. As example, one might have a file of structures and infrared data, and wish to examine possible correlations of a given substructure with infrared peaks. The hit record containing the substructure search results can be examined sequentially against several different infrared peaks using several cascaded data searches.

Results from searches do not appear on the screen immediately, but must be called out in a specific display step. It is usually considered desirable to display structures as they are encountered in the search because, psychologically, this makes the search appear to go faster and relieves the tedium of waiting. Searches in chemLit and chemFile are reasonably fast, so a separate display step is practical. A demonstration database with 196 records occupying 80% of the allocated disk space is searched in 110 seconds, and most of this is disk access time; a hard disk would reduce search time to about 40 seconds. The advantage of a separate display step is that for long searches the user can walk away and let the computer do its work. With dynamic display, the user must attend the machine and prompt it to resume searching after each display.

The Main Menu

The main chemFile menu appears in Figure 7 and contains some features other than structure entry and database searching (Options 2 and 3) to be discussed now. The main menu for chemLit offers these same features, and does not need further description.

The DISPLAY option (number 5) provides database maintenance facilities. A record in the database is retrieved by name (or substring). The structure and data in the record are displayed. DISPLAY provides file maintenance facilities for sending

```
                    S E A R C H

                  (MOUSE Control)


   1 Search for substructure

   2 Search for data

   3 Restore/Save search results

   4 Use results of last search; search for substructure

   5 Use results of last search; search for data

   6 Display search results

   7 Generate a printed report or data file

   8 Quit; return to chemFile menu

   Select number. Press ENTER to execute, or F1 for help.
```

Figure 6. The main search menu.

```
                        c h e m F i l e


    1 System setup

    2 Add structures and data to the database    (MOUSE Control)

    3 Search database

    4 List structure names in the database

    5 Display a structure, edit data or delete the record

    6 Display field definitions and file status

    7 Quit chemFile; return to monitor

    Select number followed by ENTER to execute, or F1 for help


    copyright (c) 1985, COMPress                    Dr. John Figueras
    All rights reserved.  (Version 1.2)                       Author
```

Figure 7. The main menu, chemFile.

the record to a printer, for deleting the record, or for editing (i.e., changing) data (chemFile) or text (chemLit). Deleted records are flagged but remain on the disk. The space they occupy can be recovered by a utility called COMPRESS, available from the SETUP menu (reached from Option 1 on the main menu).

The LIST option (number 4) provides an alphabetized listing of the names of the records in the database. This is useful to have in conjunction with the DISPLAY option, or the GET option in structure entry.

Characteristic of chemFile (but not chemLit) is display of the data field definitions, option 6. These definitions are installed when the database is established using the initializing procedures available from the SETUP menu (entered from Option 1 on the main menu).

The SETUP option provides access to utilities that deal with file initialization and the computer hardware environment. Through SETUP, one can do the following things:

1. Select the drive (floppy or hard disk/subdirectory) that contains the database.
2. Initialize the database file.
3. Select the proper mouse name (which varies from one manufacturer to another).
4. Enable a pen plotter and provide proper serial interface parameters for the device.
5. Remove files flagged as deleted and recover disk space.
6. In the case of chemFile, set up the file of data and field definitions.

Suggested Applications

These programs were developed for the purpose of allowing chemists to establish their own personal databases for literature searches and research results. The programs may be used for stockroom inventory (chemFile with its editing capabilities would be useful here), reaction indexing (chemLit), and structure/data correlations (chemFile). A recently developed utility, TRANSFER, permits the records extracted by a search to be collected into one file on a floppy disk. The file can be appended to a chemLit (or chemFile) database. This offers a novel means by which chemists can exchange chemical information.

Acknowledgments

I want to thank Dr. Fred Clough of COMPress, Wentworth, NH, and Dr. Duane Parker of Dynamac Corporation, Rockville, MD, for their valuable help in uncovering bugs and suggesting design improvements through a long period of debugging these very complex programs.

Literature Cited

1) G. von Kiedowski and A. Eifert, <u>Intelligent Inst. and Comp.</u>, 1986, <u>4</u>, 110-113.
2) M. L. Contreras, M. Deliz, A. Galaz, R. Rozas;, and N. Sepulveda, <u>J. Chem. Inf. Comput. Sci.</u>, 1986, <u>26</u>, 105-8.

RECEIVED March 11, 1986

Chapter 6

Applications of Personal Computer Products for Chemical Data Management in the Chemist's Workstation

Donna del Rey

Molecular Design Limited, 2132 Farallon Drive, San Leandro, CA 94577

The Chemist's Personal Software Series is a family of personal computer products designed to enhance the chemist's productivity in managing chemical information - chemical structures, reactions and data - and in writing reports which integrate graphic images with text. Examples of applications demonstrating a variety of solutions and improvements to problems regarding workstation data management are provided.

The IBM Personal Computer (PC) is becoming the chemical industry standard for workstation computers. To remain current with the move to workstation automation, and to continue to provide the chemical industry with structure-integrated programs, Molecular Design Limited has developed the Chemist's Personal Software Series. This family of professional software products is designed to assist individual chemists in their daily work and to increase their efficiency. The series consists of four programs: ChemBase, ChemText, ChemTalk and ChemHost.

Each of these programs fills a specific need defined by the chemical community. ChemBase, the PC-based chemical database management system, enables the chemist to store and search structures, reactions and textual data, and to tailor the format and content of chemical data reports on screen and on paper. ChemText, the chemical word processor, allows the preparation and printing of documents integrating graphic images, including molecules, reactions, forms and formulas, with text. ChemTalk and ChemHost work together to fulfill the PC-mainframe communication needs, including terminal emulation, error-free file transfer and streamlined access to databases created using Molecular Design Limited's MACCS software. Each PC-based program can either be used alone, or in conjunction with each other or with other programs.

Together the four programs provide a complete workstation environment for the storage, retrieval, reporting and communication of chemical information. To illustrate these features, data from an article in the *Journal of Medicinal Chemistry* (1) are used. The article describes the syntheses and testing of a series of compounds **1** expected to have antiinflammatory and analgesic activity based on previous results with derivatives of type **2**.

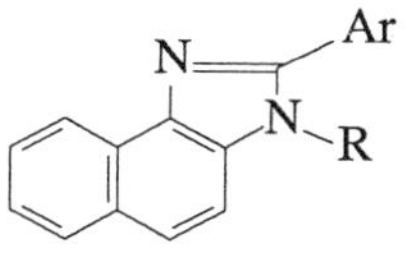

1 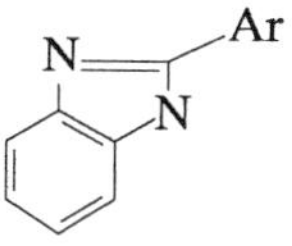   **2**

The data found in the published article will be used for the following applications of the Chemist's Personal Software Series:

- ChemTalk terminal emulation to run Molecular Design Limited's reaction indexing software REACCS on a mini/mainframe computer for access to centralized structure-based synthetic information;
- ChemBase data sorting to organize compounds by activity;
- ChemBase structure/data manipulation for structure-activity correlation;
- ChemBase data searching to identify compounds with favorable test results;
- ChemBase reaction/structure searching to streamline traditional file card methods;
- ChemTalk direct access to MACCS databases to retrieve centralized information regarding commercially available compounds; and
- ChemText document preparation to create a status report of the project results.

Applications

The chemist proposed the reaction scheme (Scheme I) for synthesizing the desired compounds. The first task is to find detailed synthetic data for the individual steps by using the terminal emulator capability of ChemTalk to search mini/mainframe reaction databases. ChemTalk provides VT100 terminal emulation for non-graphic communication and a special graphics terminal emulator for Molecular Design Limited programs. It also allows access to on-line services like CAS with Tektronix 4010 emulation. The ChemTalk graphics terminal is used here to run the mini/mainframe program REACCS to search for the preparative details of a specific step in the reaction scheme. Figure 1 shows one of the search hits as seen on a REACCS screen, describing the transformation of naphthols to naphthylamines. Similarly, reaction conditions for all steps in Scheme I were found using the PC as a high resolution graphics terminal running REACCS.

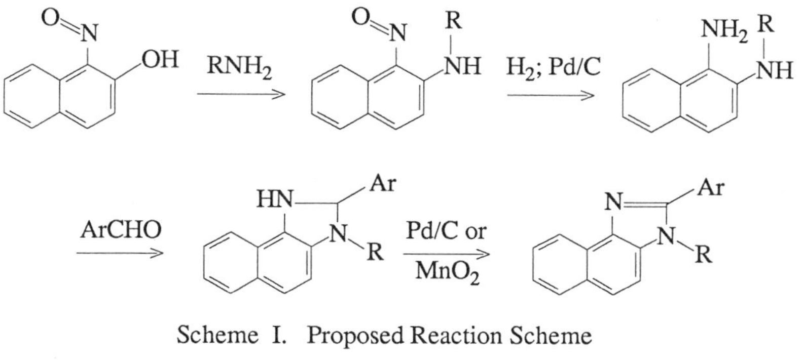

Scheme I. Proposed Reaction Scheme

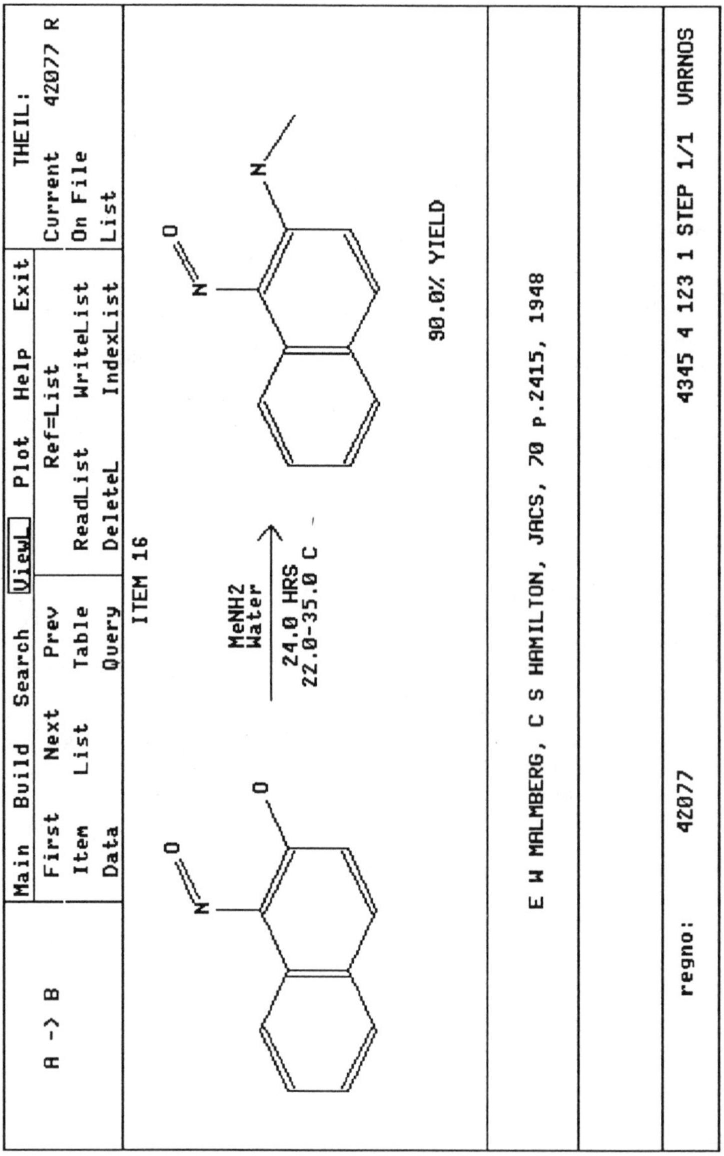

Figure 1. REACCS Search Result

Utilizing the information thus obtained, several derivatives of **1** were synthesized and subsequently tested. The following information is added to a ChemBase database dedicated to the project: structures, melting points, and LD_{50} and CE values (lethal dosage and carrageenin-induced paw edema primary testing information). ChemBase automatically calculates the formula and molecular weight when the structure is drawn.

Data for the first set of compounds, consisting of twenty-five 2-aryl-derivatives, are shown in Figure 2. This ChemBase spreadsheet format shows compounds and associated data in the order they were entered.   By viewing multiple entries, comparison of information is simple.   ChemBase also provides the ability to sort a list of compounds in ascending or descending order by data fields.   This feature is illustrated in Figure 3 in which the compounds are organized by descending order to rank them according to activity, allowing investigation of structure-activity correlations.

Inspection of the structures for the entries in the table shows that antiinflammatory activity is increased by strong electron-donating groups in the para position of the aryl substituent, while electron-withdrawing substituents show much less activity (Scheme II).

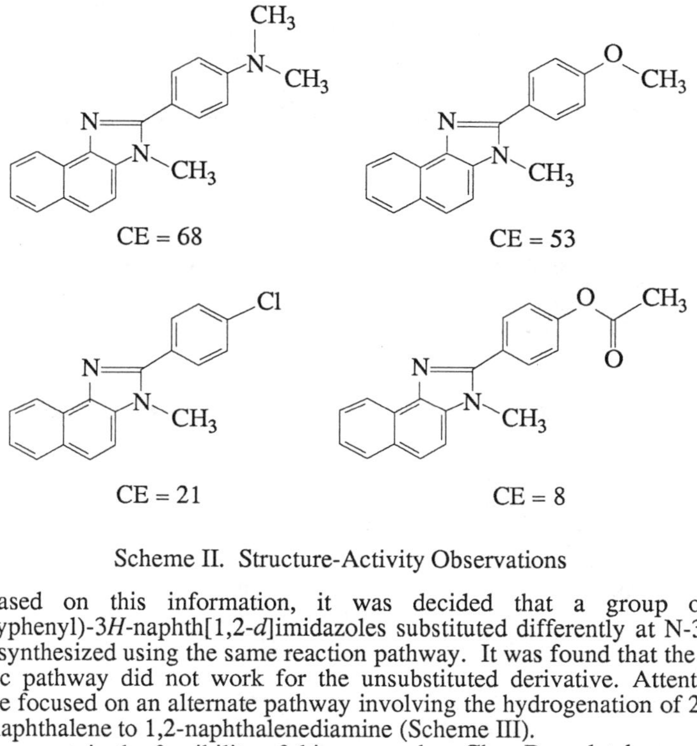

Scheme II. Structure-Activity Observations

Based on this information, it was decided that a group of 2-(4-methoxyphenyl)-3*H*-naphth[1,2-*d*]imidazoles substituted differently at N-3 should also be synthesized using the same reaction pathway. It was found that the original synthetic pathway did not work for the unsubstituted derivative. Attention was therefore focused on an alternate pathway involving the hydrogenation of 2-amino-1-nitronaphthalene to 1,2-naphthalenediamine (Scheme III).

To ascertain the feasibility of this approach, a ChemBase database consisting of hydrogenation reactions accumulated over years at the research site was searched for similar or identical reductions in order to obtain reaction conditions

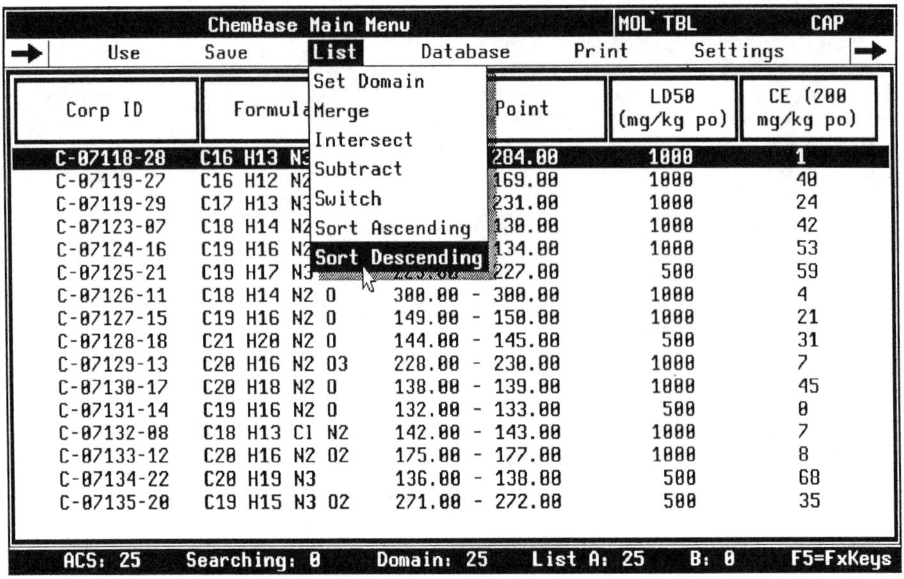

ChemBase Main Menu — MOL TBL — CAP

Use | Save | List | Database | Print | Settings

Menu (List dropdown):
- Set Domain
- Merge
- Intersect
- Subtract
- Switch
- Sort Ascending
- Sort Descending

Corp ID	Formula	Point	LD50 (mg/kg po)	CE (200 mg/kg po)
C-07118-28	C16 H13 N3	204.00	1000	1
C-07119-27	C16 H12 N2	169.00	1000	40
C-07119-29	C17 H13 N3	231.00	1000	24
C-07123-07	C18 H14 N2	130.00	1000	42
C-07124-16	C19 H16 N2	134.00	1000	53
C-07125-21	C19 H17 N3	227.00	500	59
C-07126-11	C18 H14 N2 O	300.00 - 300.00	1000	4
C-07127-15	C19 H16 N2 O	149.00 - 150.00	1000	21
C-07128-18	C21 H20 N2 O	144.00 - 145.00	500	31
C-07129-13	C20 H16 N2 O3	228.00 - 230.00	1000	7
C-07130-17	C20 H18 N2 O	138.00 - 139.00	1000	45
C-07131-14	C19 H16 N2 O	132.00 - 133.00	500	8
C-07132-08	C18 H13 Cl N2	142.00 - 143.00	1000	7
C-07133-12	C20 H16 N2 O2	175.00 - 177.00	1000	8
C-07134-22	C20 H19 N3	136.00 - 138.00	500	68
C-07135-20	C19 H15 N3 O2	271.00 - 272.00	500	35

ACS: 25 | Searching: 0 | Domain: 25 | List A: 25 | B: 0 | F5=FxKeys

Figure 2. Descending Order Sort on a ChemBase Database

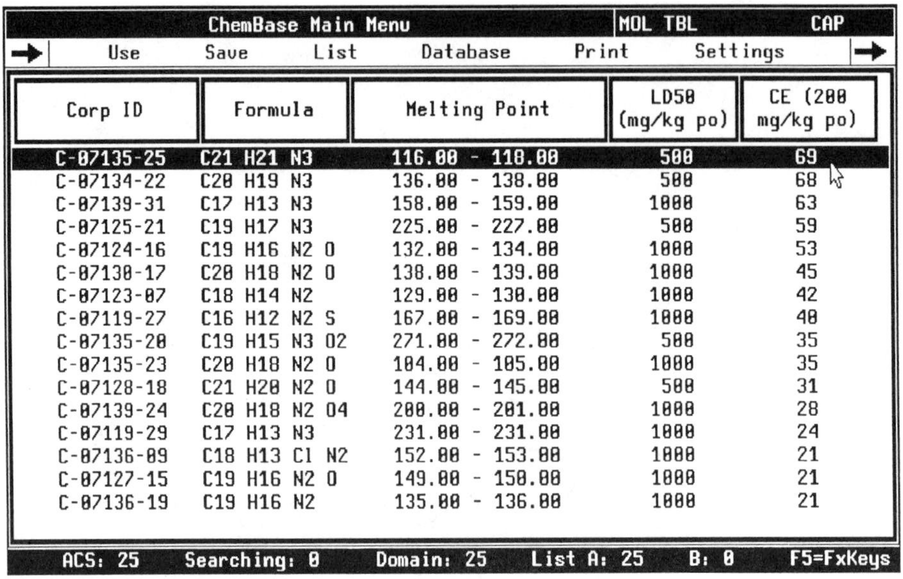

ChemBase Main Menu — MOL TBL — CAP

Use | Save | List | Database | Print | Settings

Corp ID	Formula	Melting Point	LD50 (mg/kg po)	CE (200 mg/kg po)
C-07135-25	C21 H21 N3	116.00 - 118.00	500	69
C-07134-22	C20 H19 N3	136.00 - 138.00	500	68
C-07139-31	C17 H13 N3	158.00 - 159.00	1000	63
C-07125-21	C19 H17 N3	225.00 - 227.00	500	59
C-07124-16	C19 H16 N2 O	132.00 - 134.00	1000	53
C-07130-17	C20 H18 N2 O	138.00 - 139.00	1000	45
C-07123-07	C18 H14 N2	129.00 - 130.00	1000	42
C-07119-27	C16 H12 N2 S	167.00 - 169.00	1000	40
C-07135-20	C19 H15 N3 O2	271.00 - 272.00	500	35
C-07135-23	C20 H18 N2 O	104.00 - 105.00	1000	35
C-07128-18	C21 H20 N2 O	144.00 - 145.00	500	31
C-07139-24	C20 H18 N2 O4	200.00 - 201.00	1000	28
C-07119-29	C17 H13 N3	231.00 - 231.00	1000	24
C-07136-09	C18 H13 Cl N2	152.00 - 153.00	1000	21
C-07127-15	C19 H16 N2 O	149.00 - 150.00	1000	21
C-07136-19	C19 H16 N2	135.00 - 136.00	1000	21

ACS: 25 | Searching: 0 | Domain: 25 | List A: 25 | B: 0 | F5=FxKeys

Figure 3. Sorted Database in Tabular Format

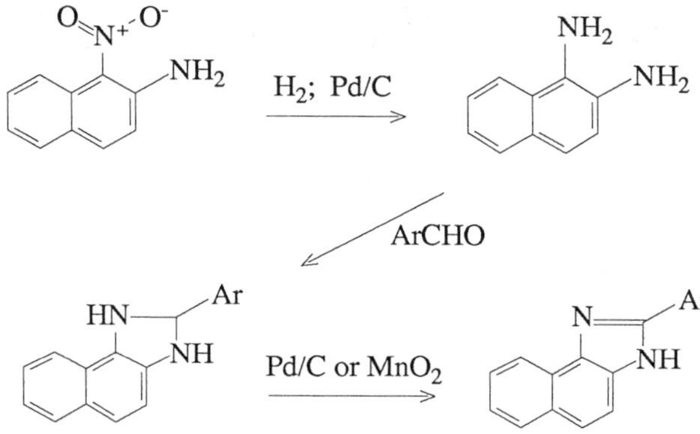

Scheme III. Alternate Synthetic Pathway for Unsubstituted Compounds.

for this particular reduction. The differences between this example of structural searching in ChemBase and the earlier REACCS searching example are conceptual: REACCS databases reside on a mini/mainframe computer as a centralized source of synthetic information, while PC-based ChemBase databases are tailored to specific projects or reaction topics. The Transformation Substructure Search (TSS) query employed in this ChemBase search is shown in Figure 4. Figure 5 shows one of the TSS hits, including examples of reaction-specific data fields.

This use of a ChemBase reaction database containing proprietary hydrogenation information is a good example of how a ChemBase reaction database can replace a traditional file card system. The ChemBase database is more easily accessed and can be duplicated for other users.

Next, there was a need to ascertain whether the starting material for the required hydrogenation was commercially available. A previously created form (Figure 6) designed to show supplier information was retrieved, and the nitro derivative was entered into it. By exiting directly from ChemBase to ChemTalk, this display and all the information in it was preserved. The structural image could now be used to perform a search of the Fine Chemicals Directory (FCD), a mini/-mainframe database containing information supplied by Fraser-Williams on commercially available compounds. Using the ability to search MACCS mainframe databases directly from ChemTalk, the chemist can search for an identical match by using the Find Current command (Figure 7). Figure 8 displays the structure with its supplier information from FCD.

With the information on the hydrogenation reaction and the availability of the starting material in hand, the unsubstituted derivative was synthesized. This compound along with the other newly synthesized compounds was then added to the existing database with their respective data. At this point, it was important to determine which compounds showed the most potential so that they could be further tested. A data search minimizing toxicity and maximizing anti-inflammatory data (LD_{50} values greater than or equal to 100 and CE values greater than 40) was performed (Figure 9). Data for the compounds that met these criteria were automatically displayed in the current table format (Figure 10). Secondary testing was done on these promising compounds, and fields for these new tests were added to the project ChemBase database. An example of one of the active compounds is shown in Figure 11 with secondary test results.

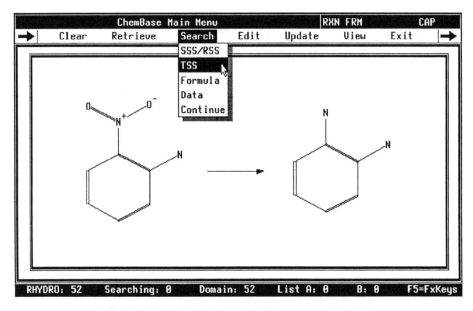

Figure 4. ChemBase Transformation Substructure Search Query

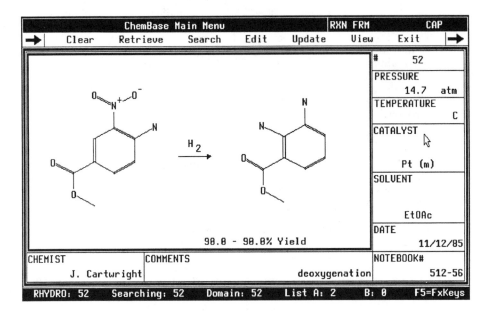

Figure 5. Transformation Substructure Search Result

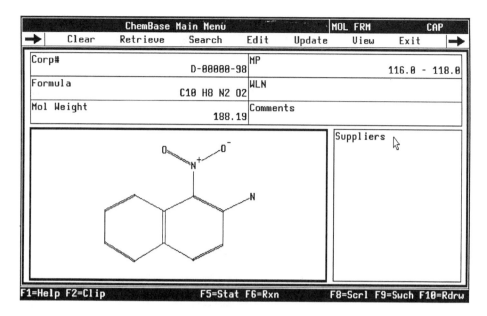

Figure 6. ChemBase Form for Starting Material

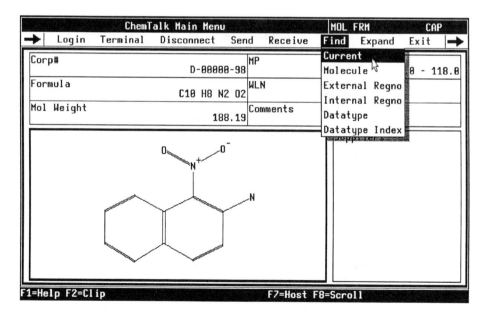

Figure 7. ChemTalk Command Execution

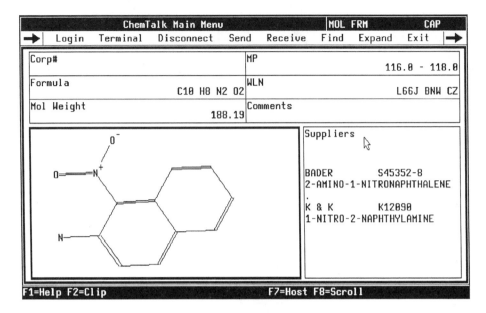

Figure 8. FCD Find Current Search Result in ChemTalk

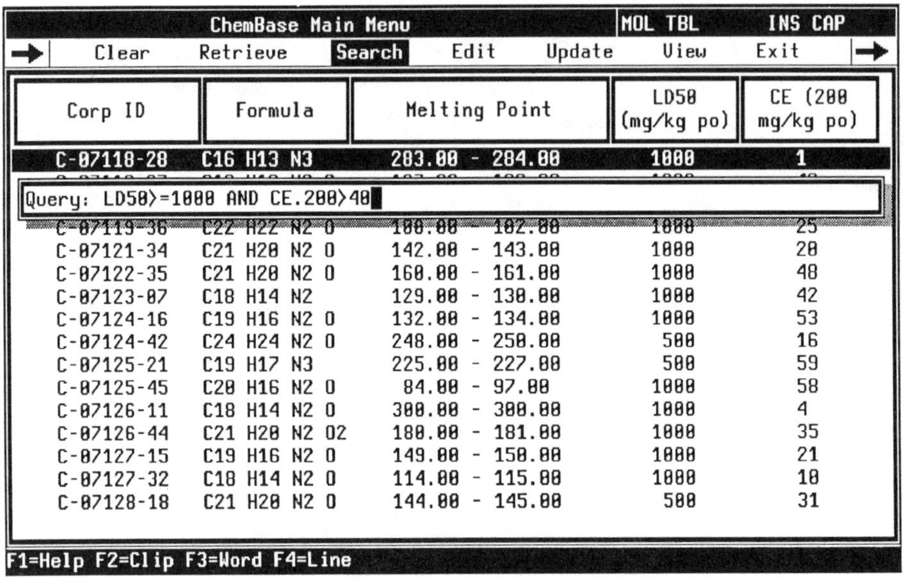

Figure 9. Multiple Field Data Search Query

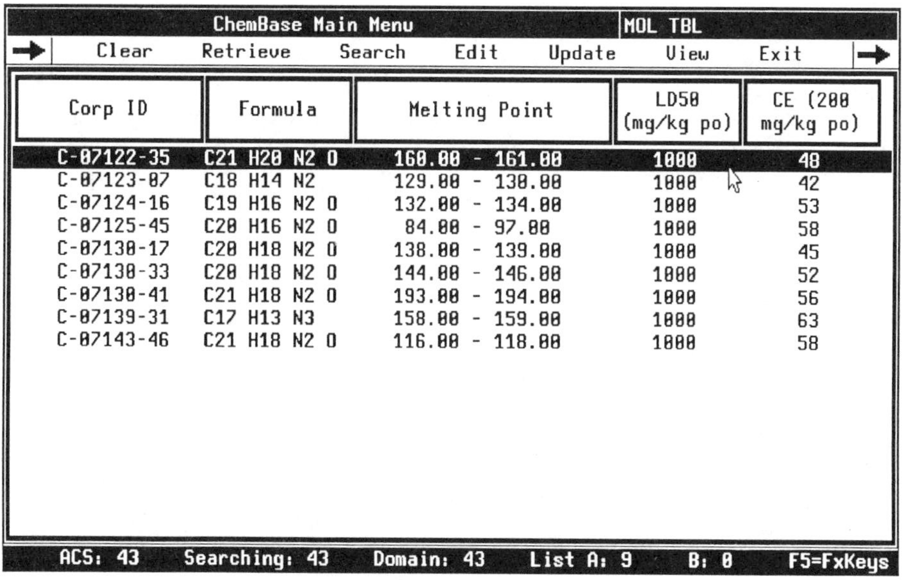

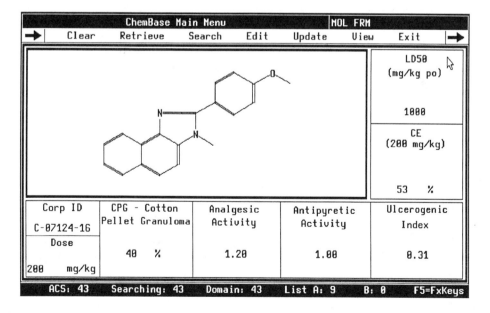

ChemBase Main Menu
MOL TBL
Clear
Retrieve
Search
Edit
Update
View
Exit
Corp ID
Formula
Melting Point
LD50 (mg/kg po)
CE (200 mg/kg po)
C-07122-35 C21 H20 N2 O 160.00 - 161.00 1000 40
C-07123-07 C18 H14 N2 129.00 - 130.00 1000 42
C-07124-16 C19 H16 N2 O 132.00 - 134.00 1000 53
C-07125-45 C20 H16 N2 O 84.00 - 97.00 1000 58
C-07130-17 C20 H18 N2 O 138.00 - 139.00 1000 45
C-07130-33 C20 H18 N2 O 144.00 - 146.00 1000 52
C-07130-41 C21 H18 N2 O 193.00 - 194.00 1000 56
C-07139-31 C17 H13 N3 158.00 - 159.00 1000 63
C-07143-46 C21 H18 N2 O 116.00 - 118.00 1000 58
ACS: 43
Searching: 43
Domain: 43
List A: 9
B: 0
F5=FxKeys

Figure 10. Data Search Result

ChemBase Main Menu
MOL FRM
Clear
Retrieve
Search
Edit
Update
View
Exit
O
N
N
LD50 (mg/kg po)
1000
CE (200 mg/kg)
53 %
Corp ID
C-07124-16
Dose
200 mg/kg
CPG - Cotton Pellet Granuloma
40 %
Analgesic Activity
1.20
Antipyretic Activity
1.00
Ulcerogenic Index
0.31
ACS: 43
Searching: 43
Domain: 43
List A: 9
B: 0
F5=FxKeys

Figure 11. Structure/Data Display

At this point all the preliminary testing was completed. To find compounds that show a better pharmacological profile than control compounds, a new search was carried out over the last set of nine compounds. The data search query is shown in Figure 12. The result of this search was automatically displayed in the current form (Figure 13).

At this point in the project, a status report of the results obtained thus far was required. The report needed to include structures of interest, tables of test results, reaction schemes and other chemical information. This information can be integrated into a document using ChemText, the chemical word processor. Not only does ChemText provide the word processing capabilities necessary to write complex reports, but also allows integration of information that was previously generated in ChemBase. All additional drawing or notating can be done in a sketching area within ChemText itself. Figure 14 illustrates structures integrated with text explaining structure-activity relationships. Figure 15 shows a table that was generated in ChemBase and, as an ASCII file, was moved into ChemText in order to add the subscripts, lines and other aesthetics. The last figure (16) shows part of the reaction scheme that has been placed in a section of the status report explaining the synthesis of the compounds.

ChemText, in this example, has been used to create a document utilizing images and data that already exist in ChemBase. ChemText provides features common to chemical documents: scientific, foreign and mathematical fonts, italics, boldface, sub- and super-scripts, and general graphic sketch features, allowing the chemist flexibility in document creation.

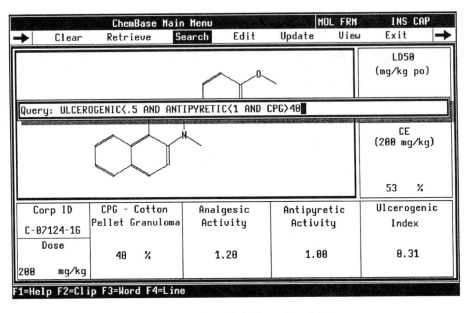

Figure 12. Multiple Field Data Search Query

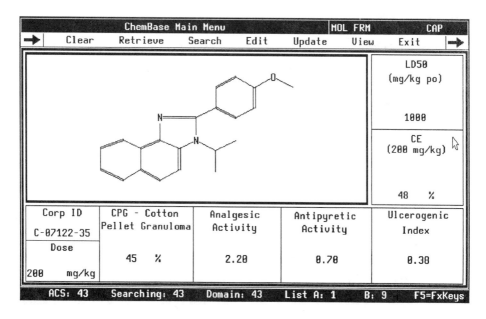

Figure 13. Data Search Result

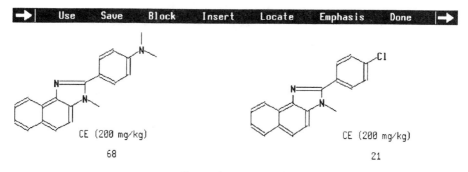

Inspection of the table shows that antiinflammatory activity is increased by strong electron-donating groups in the para position of the aryl substituent, while electron-withdrawing substituents show much less activity (Figure 4).

Figure 14. ChemText Structures and Text

Biological Results and Discussion

The test compounds were screened for antiinflammatory activity by means of the carrageenin-induced rat paw edema test (CE), and the percentages of inhibition relative to the controls are presented in Tables II and III.

Table II. Chemical Data and Carrageenin Edema Inhibition

No.	R	mp, °C	LD_{50} (mice) mg/kg po	CE 200 mg/kg po
7	C_6H_5	129.00 - 130.00	1000	42
8	$3-ClC_6H_4$	142.00 - 143.00	1000	7
9	$4-ClC_6H_4$	152.00 - 153.00	1000	21
10	$2-HOC_6H_4$	143.00 - 144.00	500	1
11	$4-HOC_6H_4$	300.00 - 300.00	1000	4
12	$4-AcOC_6H_4$	175.00 - 177.00	1000	8

```
----+----L----+----2----+----3----+----4----+----5----+----6----+----7----+--R-T8----+-T--9
1:PS       Ln:31  Col:77  Roman              1      F5=STAT      INS    CAP
```

Figure 15. ChemText Chemical Table

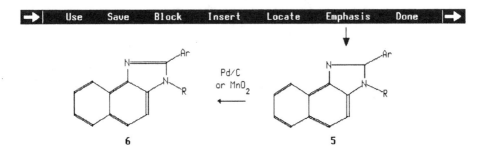

After hydrogenation condensation of the resulting 2-alkyl-1,2-naphthalenediamines (**4**) with substituted benzaldehydes afforded the 1,2-dihydro derivates (**5**), which were converted to the final compounds (**6**) by means of palladium on carbon (method A) or manganese dioxide (method B). Dehydrogenation with palladium gave higher yields and cleaner reaction products and allowed the conversion of **3** to **1** to be done in "one pot". The high yields and the mild reaction conditions of the first step make **2** an ideal starting material for the synthesis of diamines **4**. However, *tert*-butylamine failed to react with **2**; thus, 3-*tert*-butyl-2-(4-methoxyphenyl)-3*H*-naphth[1,2-*d*]imidazole is prepared from 2-(*tert*-butylamino)-1-nitronaphthalene

```
----L----T----+----2----+----3----+----4----+----5----+----6----+----7----+---T8----R-T--9
1:PS       Ln:11  Col:83  Roman              1      F5=STAT      INS
```

Figure 16. ChemText Reaction, Data and Text

Conclusion

Information from a publication in the *Journal of Medicinal Chemistry* has been used to illustrate some potential applications of the Chemist's Personal Software Series, ranging from chemical database management to mini/mainframe database access to report generation. ChemBase, ChemTalk, ChemHost and ChemText work together to provide a complete workstation information management system.

Acknowledgments

The author is grateful to P. Cohan for assistance with application review, and to Dr. G. Grethe for helpful discussions regarding scientific ideas and manuscript review.

This paper was done entirely using ChemText and printed on an Apple LaserWriter.

Literature Cited

1. Toja, E.; Selva, D.; Schiatti, P. *J. Med. Chem.* **1984**, *27*, 610-616.

RECEIVED March 11, 1986

Chapter 7

Design and Development of an Interactive Chemical Structure Editor

Joseph R. McDaniel and Alvin E. Fein

Fein-Marquart Associates, Inc., 7215 York Road, Baltimore, MD 21212

The US National Cancer Institute needed a method for
easily and quickly entering large numbers of chemical
structures into a database. Command-driven systems and
Teletype systems driven by keyboard input had been tried,
but these systems were judged to be clumsy, difficult to
learn, and slower than a graphics system should be. The
result was the design and development of a
microcomputer-oriented program which has evolved into a
product known today as SuperStructure. This paper
discusses the design criteria and the development
decisions which resulted in a fast and easy-to-learn
chemical structure editor.

The initial problem in this project involved the hardware
configuration. Previous experience with "dumb" terminal graphics
systems operating over telephone lines at 120 characters per second
had indicated that the responsiveness of such systems would never be
adequate in terms of both entry throughput and user patience:
unintelligent terminals controlled by a central processor would
require an inordinate number of data-input personnel because of the
slowness of data throughput and because of poor productivity arising
from frustration with lack of system responsiveness.
    To avoid this bottleneck, it was decided to design a
distributed computing system: the actual entry and editing of
chemical structures would be done on a local microcomputer that was
connected to the mainframe system only for uploading and downloading
the data needed to reproduce the desired chemical structure. The
local microcomputer system would operate as a normal "dumb" terminal
when being used as a data-entry and query device against the
mainframe; it would operate as a stand-alone processor for the
graphic input or editing of individual chemical structures.

Selection of the Hardware

The next step was to examine available microcomputer systems to find
one suitable for the remote entry of chemical structures.

0097-6156/87/0341-0062$06.00/0
© 1987 American Chemical Society

Obviously, one key criterion was that the microcomputer have good graphics capability. This entailed determining what would be necessary for entry of chemical structures of up to 255 atoms. (The upper limit of 255 atoms was chosen primarily because Chemical Abstracts Service structures are truncated to 255 atoms.) After examining several classes of compounds, we determined that the height and width of the structure with bonds and atom symbols would be controlled largely by two requirements:

1. The need to display an atom symbol of two characters (along with optional mass and charge information).
2. The need to display bonds long enough for the different bond type symbols (single line for single bond, double line for double, etc.) to be readily identifiable when viewed on the local microcomputer screen.

Also, it was noted that large structures, when drawn for publication, tend to be relatively compact, with heights and widths that are approximately equal. In cases where the structure is very "long," the publication technique is to fold the structure, a technique which was adjudged acceptable in our case as well. Thus, structures tended to be no larger than about 20 atoms high by 20 atoms in width.

The height and width of a character are minimally about 8 by 8 pixels, respectively, for readability and reasonably good formation of the character. Also, the length of a line used to represent bonding should be proportional to the character height and width both for pleasing proportions in the resulting structure and for "readability," the ability to distinguish between bond symbols. Thus, it was determined by trial that the minimum pleasing bond length is about 3/2 times the height or width of the character, whichever is greater. The final estimation of resolution required was thus determined to be about $20(n + 3n/2)$ [where n is the character size in pixels] or about 400 by 400 pixels. Compromises could be made by trading off the normally better horizontal resolution of raster graphics display devices for reduced vertical resolution, but the limiting resolution could not differ too markedly without compromising ease of entry of structures by forcing operators to remodel entries to fit the screen.

Another hardware criterion was the ability to support the size of programs expected for performing terminal communications, graphics editing, and program development (since it was desirable to do the development on the same microcomputer chosen for production). At the time the initial design work was taking place (in early 1983), the choices for suitable remote microcomputer systems were primarily those based on the 8-bit Zilog Z80 and MOS Technology 6502 and the, then, relatively new 16-bit microcomputers based on the Motorola 68000, or the Intel 8088. Since the programs were obviously going to be large and complex, it was felt that a large address space would be needed, thus pointing strongly toward a 16-bit, 68000- or 8088-based system. In addition, the graphics resolution requirements of the system were beyond the graphics capabilities of the 8-bit systems then available. Thus, 8-bit systems were rejected.

Of the remaining 16-bit microcomputer systems available then, some were rejected because they would have required assembly of a complete hardware package from the wares of a variety of small suppliers (an option that was unacceptable because of maintenance difficulties and fear of unreliable support during the course of what would become a long-term project). Other 16-bit systems (including all those based on the 68000) were rejected because of inadequate integrated graphics support.

This process of elimination left two 8088-based systems in the running. One of these was the, then, recently introduced IBM PC. However, the graphics capability then available on the IBM PC was limited to 640 horizontal by 200 vertical pixels. Furthermore, the two general types of graphics input devices for the IBM PC were inadequate: light pens were severely limited as to resolution (being able to resolve basically only a character cell of 8 by 8 pixels), and mice were not well supported by software and tended to be available from relatively small, unstable companies.

Fortunately, none of these problems affected the remaining candidate, the Victor 9000. The Victor 9000 had been favorably reviewed (1) and was being produced by a subsidiary of Victor Business Products, a subsidiary of Kidde, Inc. The superior resolution of 800 horizontal by 400 vertical pixels for the Victor 9000 also was very favorable and fit the calculated needs quite well. In addition, the sytem had standard graphics capability and communications support, 1.2 megabyte diskettes, and suitable development tools such as a FORTRAN compiler and a Graphics Tool Kit to simplify the graphics programming. Finally, although not manufactured by Victor Technologies, a pointing device known as a TOUCHPEN (trademark of Sun-Flex Company, Inc., 20 Pimentel Court, Novato, CA 94947) was supported, and this proved to be superior to a light pen in terms of resolution, sensitivity, and reliability of operation. Consequently, the Victor 9000 was chosen.

Initial Program Development

After we had acquired the hardware, our first step in program development was to determine the mode of interface with the user. Several small programs were created to model various user interfaces focusing on whether input should be largely via the pointing device or largely via the keyboard. It was determined that it was faster and more user-oriented to keep all interactions on the screen with selection via the TOUCHPEN. This reduced the need for typing skill and the need to shift one's focus from the screen to the keyboard frequently. A menu positioned along the right side of the screen was chosen as the means of relating the functions of the program to the needs of the user. Positioning on the right side of the screen was determined by the need to preserve vertical pixels for structures (as the horizontal requirements exceeded those needed by a factor of two while the vertical resolution was essentially equal to that needed). Only a small area of the bottom of the screen was set aside to be used for messages and sub-menus, the area occupying only 16 pixels. (Characters are 10 pixels wide and 16 pixels high on the Victor 9000.)

This early work used the FORTRAN compiler and Graphics Tool Kit supplied by Victor. In both cases there were severe programming limitations. The Graphics Tool Kit was very slow since its approach was that of an ANSI screen driver, responding to character strings sent to the screen and performing graphics operations rather than

displaying the characters directly. The maximum data rate of the
screen was approximately 900 characters per second; consequently, it
was difficult to send a series of commands quickly enough to the
screen to achieve an interactive user interface. The FORTRAN
compiler (supplied by Microsoft Corporation to Victor) was also very
slow, required huge amounts of disk space for intermediate files
during compilation, was a minimal FORTRAN 77 implementation, and
contained many errors. Even with these limitations, the system was
usable, and initial development proceeded quite well once methods
had been discovered to work around the problems.

One example of methods developed to work around software
limitations involved the requirement for a menu of cyclic nuclei as
an aid in constructing structures. A partial solution to the
slowness of the Graphics Tool Kit was to take advantage of the
capacity for multiple screens (with only one visible at a time).
Thus the menu of cyclic nuclei was placed on a separate screen which
could be "popped" into view almost instantly. In fact, we could
alternate among several screens quickly without rewriting. Thus the
early versions of the system tended to have several different
screens for operations so as to minimize operator wait time. This
approach was not optimal, but was dictated by the nature of the
tools available at the time.

Initial Design Criteria to Test the Concept

The initial capabilities which were implemented to prove the
approach included:
1. The ability to select from a standard ring system menu to
   initiate the structure. (The initial version only allowed this
   once at the beginning of structure creation.)
2. The ability to add atoms and bonds to a structure by moving the
   TOUCHPEN from point to point on the screen. Such points can be
   added in an optional mode which permits placement of atoms only
   on horizontal, vertical, or 45-degree diagonals with respect to
   the previous atom to force structural protocol.
3. The ability to specify particular atom types and bonds at any
   point in the structure.
4. The ability to delete particular atoms or bonds from the
   structure.
5. The ability to create a connection table corresponding to the
   structure being constructed.
6. The ability to produce a printed copy of both the structural
   diagram and the corresponding connection table.
Admittedly, this set of capabilities was fairly minimal, but it
was sufficient to demonstrate the ideas involved and to permit
experimentation with the system to see what concepts needed to be
changed. Also, the only communication capability initially
available was to save the structure on disk and then transfer the
information from disk to the mainframe system in some manner.

Initial Implementation Decisions

The operation of the system was almost completely driven by using
the TOUCHPEN to select a menu item from a box on the screen or by
touching a place on the screen where an atom was to be drawn,

deleted, or edited. However, the TOUCHPEN software permits detection of both initial contact with the screen as well as the breaking of contact. By experimentation, it was determined that the more natural operation mode was to perform an operation upon the lifting of the pen rather than upon initial contact. The problem with operating on initial contact was that the users tended to "stab" at the screen and often touched the wrong spot. With the software responding only to the lifting of the pen, it was possible for the user to reposition the pen to the desired spot (as indicated by a cursor display on the screen) before initiating the operation.

A cursor was used both to aid in indicating where the TOUCHPEN was placed and to indicate where atoms would actually appear. This was particularly useful in the mode which limited atom placement at 45-degree multiples with respect to the previous atom drawn. To draw a chain or ring, the user would select the DRAW mode (by touching its menu box) and place the initial atom anywhere in the structure portion of the screen. Subsequent atoms would then be forced into positions on 45-degree multiple radials from the previous atom drawn until the DRAW mode was again selected (after which the first atom drawn would be free of the angular restriction with regard to any previous atom unless the first location selected already contained an atom symbol).

All atoms were drawn with whatever atom symbol had last been selected. To select an atom symbol the user touched the LABEL ATOM menu item, which caused a secondary menu of atom symbols to appear at the bottom of the screen in the message area previously reserved. The list was chosen by frequency of appearance in organic compounds as determined by an examination of compounds in the Structure and Nomenclature Search System (a component of The Chemical Information System). Similarly, bonds were drawn using the last selected bond type.

The connection table consisted of the normal atom symbol and bonding information fields and included extra fields for atom charges and masses as well as the (x,y) coordinates for each atom. Additional information was present for the strings of characters used for molecular formulas, for dot disconnect moieties, and for ill-defined structures where only the formula was known for a portion of the structure.

To reduce the overhead of both internal operations to maintain the connection table and the transmission time for sending the table to the mainframe, only the upper left triangle of the connection information was kept: the lower right portion could always be regenerated whenever needed. (If one knows that atom A is connected to atom B, one also knows that atom B is connected to atom A even if the latter statement is not explicitly retained.) The only disadvantage of keeping only one-half of the connection information occurred during deletions of atoms and other operations which needed the connecting links both to and from an atom. In those cases direct indexing was impossible and a search was necessary. The advantages of not having to renumber, insert, and delete the extraneous information more than compensated for the extra work necessary on those few occasions where the reverse information was necessary.

Initial User Response

After reviewing the initial system, the user community requested
several changes. One was creation of a grid of dots (pixels) to
serve as guide points for the creation of new atoms. Even though
the system had been designed to force all bond angles to be
multiples of 45 degrees, the distance between atoms was still
variable, and the guide dots would aid in placement of additional
atoms.

The mode of labeling atoms and bonds was found to be difficult
in practice and, rather than continue with the last-selected atom
and bond types for drawing new atoms and connections, we reset the
atom type automatically so that the Luhn dot (for a cyclic carbon
atom) would be drawn by default and all initial bonds would be
single bonds by default. To change any of these the user could
select the LABEL ATOM mode to relabel an atom node or the LABEL BOND
mode to change bonding types. The relabeling of atoms was further
changed in the final development of the system, as will be explained
later.

Users also found the strain of keeping the arm elevated to
position the TOUCHPEN to be onerous. Consequently, they decided to
construct a special table with a cutout for the microcomptuer
display so that the screen would be recessed horizontally in the
table top and the user's arm could then rest almost entirely on the
table.

With only minor modifications to the initial design, further
development was begun as the users quickly recognized the advantages
of a graphics input mode and the speed achieved by being divorced
from both the vagaries of mainframe computer loading and the
uncertainties and slowness of telephone line data transfer.

Continued Development Goals

The continued development of the system addressed the following
goals:
1. Provide a complete menu system for full operation of the
   system.
2. Provide bond and atom symbol menus.
3. Expand the ring menus and allow for their use at any time, not
   just upon initial construction of a structure.
4. Create a new acyclic group menu with fragments to be attached
   to structures being created.
5. Allow for the creation of "dot disconnects" with automatic
   creation of the structure for the most common salts and acids.
6. Allow the structural diagrams to be translated, rotated, and
   mirrored.
7. Provide an ability to copy structural moieties.
8. Allow for user-defined inclusion of common nuclei or
   structures.
9. Allow for charged atoms as part of the atom symbol.
10. Implement a capability for atoms with abnormal masses.
11. Expand the bond menu to include stereo bonds.
12. Allow for poorly defined structures where portions of the
    structure are known only by molecular formula and other
    portions may have unknown attachment points.

13. Implement the terminal interface so that access to the
     structure editing package would be controlled by the mainframe
     (host) with no direct user interaction.
14. Allow for the display of information from sources expecting a
     Tektronix 4010 terminal for display, i.e., implement a
     Tektronix 4010 emulator. Again, this was to be transparent to
     the user and to require no explicit action on the user's part
     to access this mode.

Further Design Decisions

The next version of the structure editor system developed with few
major changes in concept. The terminal interface was created as a
"dumb" terminal emulator with no special formatting capabilities. A
Tektronix 4010 emulator was created with the ability to emulate only
the output features of the 4010 and none of the special input
features, such as cross-hairs, since these were never to be used.
     The interface to both the structure editing and the Tektronix
4010 emulator portions was designed to be controlled solely by the
host mainframe computer using escape character sequences. The
choice of characters was made so as to avoid conflicts with either
the Digital Equipment Corporation VT100 (since it was anticipated
that emulation of the VT100 was a future possibility) or the
Tektronix 4010 escape and control sequences.
     A protocol for transmitting and receiving structures was
implemented with a design based on a simple check-sum and packet
number to control errors. This capability thus allowed us to send
newly created structures from the Victor 9000 to the mainframe as
well as to receive previously generated structures for editing.
Rather than the instructions for drawing a structure, only a
connection table (including positional data) was transferred since
that was all that was required to regenerate the structure at either
end. The transfer algorithm sent each atom and its connection
information as a packet. Acknowledgment of each packet was required
before progression to the next one. If no acknowledgment was
returned by the receiving system within a set interval,
retransmission of the packet was automatically attempted.
     A new graphics interface was obtained from Victor which
completely bypassed the serial character interface and provided
direct memory mapped (bit mapped) updating of the screen. The
incorporation of this new software was relatively easy as the
original programming had been done by creating subroutines for the
major graphics functions (such as drawing a line, writing a string
of characters, erasing the screen, drawing a box, etc). Thus the
major changes only involved modifying the interfaces to correspond
to the new calling sequences. The effect was immediately obvious:
the new software interface functioned at least ten times faster.
The new speed of screen updating made the system extremely
efficient, whereas the previous approach had been only marginally
acceptable.

Acceptance of the Final Design

The program delivered to the National Cancer Institute consisted of
three major functional pieces: a dumb terminal interface which

controlled the other two functions via escape sequences received
from the host mainframe computer, a Tektronix 4010 emulator, and the
structure editor itself.

The structure editor had two modes which differed only
slightly: the normal mode (used to define structures to be entered
into the database system) and a query mode (for entering structures
for searching the database). In query mode, some options were
disabled (such as attachment points and ill-defined fragments).
Multivalued atom nodes and a large list of bond types were allowed
in query mode. In the same manner, in non-query mode, atoms must be
singular valued and bond types could only be single, double, triple,
stereo-up, stereo-down, or stereo-unknown.

The final structure editor routines would show the user a menu
consisting of DRAW, NORMALIZE, LABEL ATOM, LABEL BOND, DELETE ATOM,
DELETE BOND, SET CHARGE, SET MASS, DOT DISCONNECT, ATTACHMENT POINT,
MOVE MENU, TEMPLATES, PRINT, and SEND as illustrated in Figure 1.
Whatever menu option was selected would be "lit" by displaying the
text in reverse video (light background and dark letters). Once in
most modes, the system would stay there until a new mode was
selected; some modes would return automatically to the DRAW state
upon completion of an operation.

DRAW was used to put the system into the draw mode and had the
dual purpose of resetting the chaining feature so that a new chain
could be started without automatic connection to the last atom drawn
previously.

NORMALIZE controlled the 45-degree limitation of angles between
connected atoms. NORMALIZE was initially "on," and could be toggled
by touching the NORMALIZE box.

LABEL ATOM worked in conjunction with SET CHARGE and SET MASS
to relabel atom symbols. All three modes would allow a node to be
relabeled with the curently set atom symbol, mass, and charge. The
general technique would be to select the atom symbol desired (from a
sub-menu that appeared after LABEL ATOM had been touched) and then
select SET CHARGE and/or SET MASS if necessary. Finally, the user
would touch the atom(s) to be relabeled.

DELETE ATOM and DELETE BOND operated similarly. To delete
either, the user simply selected the desired mode and then touched
atoms or bonds to perform the operation. Selection of bonds was
tried using several techniques. The first was to select the bond by
extending the line and computing the distance from the TOUCHPEN
location to the line perpendicularly. This did not work well;
ambiguities were too frequent, and it was difficult to select some
bonds   e.g., if two bonds were colinear, both bonds were the same
distance from the touched point. This technique was also very
expensive computationally because of the transcendental functions
needed to compute the distance from a line. It could have been
refined by limiting the selection process so that the distance was
computed only between the end points of the bond, but this would
have been even more expensive computationally. The next trial was
to compute the radial distance from the center of the bond to the
TOUCHPEN location and select the least distance. Again, this tended
to result in too many ambiguities, and the computational costs
(involving two multiplications) were fairly high. This was refined
to reduce the "circle" of acceptance to approximately one
atom-to-atom's distance. This approach was satisfactory. The final

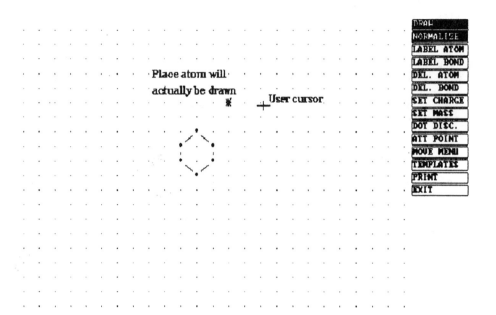

Figure 1. Structure editor menu with figure, user cursor, and
"normalized" (automatic) atom placement.

mode chosen was the simplest, and it used the sum of the x and y distances from the center of the bond to the touched point. With the same cutoff as before on distance from a bond, this worked quite well; the only problems occurred with very long bonds where the user still had to touch near the center rather than just "on the line."

The MOVE MENU bar was used to select a secondary menu list which is discussed below.

DOT DISCONNECT placed the program into a mode where a large dot was put on the screen and the user could then enter an optional molecular formula for various acids, salts, etc. If the desired formula had not been predefined, the user could simply exit, leaving the dot on the screen, and draw in the structure for the dot-disconnected moiety.

ATTACHMENT POINT was used to note (by placing an asterisk by the atom symbol) where possible attachments might occur for moieties drawn on the screen. (Any number of moieties could exist on the screen simultaneously.) Note that ATTACHMENT POINT was replaced with DISPLAY MVA (multi-valued atom) when in the query mode, which is used to display the list of possible atom symbols allowed during a search.

TEMPLATES gave the user access to a third screen which offered menus of either CYCLIC structures (Figure 2) or ACYCLIC fragments (Figure 3). These operated by displaying the appropriate "list" of structures or fragments and allowing the user to select one to be "carried" back to the structure being created or edited. Cyclic moieties could be placed anywhere on the screen as long as no conflicts existed with previous atoms or the edges of the screen. (Bond crossings were ignored and allowed to occur anywhere.) Acyclic fragments were only allowed to be attached to an existing atom node. The user would select a node and then the placement of the fragment, with rotations and mirroring being done automatically in an attempt (not always successful) to keep the diagram from having crossing bonds within the attached fragment. Checks were made to prevent coincident or off-screen atoms. Figure 4 illustrates the results of selecting a moiety from the cyclic menu and a fragment from the acyclic menu after merging them with the preexisting phenyl ring. (No claims are made for chemical correctness in these examples.)

PRINT would send a copy of the screen to an attached dot-matrix printer for a paper copy.

SEND initiated the sending of the current structure to the host computer and led to exiting from the structure editing mode.

The MOVE MENU box sent the program to a second screen with a new menu list consisting of LABEL MENU, TRANSLATE, TURN 90, TURN 180, TURN 270, MIRROR X, MIRROR Y, COPY, FRAGMENT COPY, DELETE MOIETY, DELETE ATOM, DELETE BOND, SAVE, RESTORE, and KILL. Note that DELETE ATOM and DELETE BOND were present on both menus and had the same functions in both cases. Whatever structure was present on the prior screen was copied over to the MOVE screen for further operations.

LABEL MENU was used to return to the previously described menu.

TRANSLATE allowed the user to move or translate a moiety on the screen. The user would be prompted to touch an atom in a moiety and then to touch the point to which that atom was to be translated. No

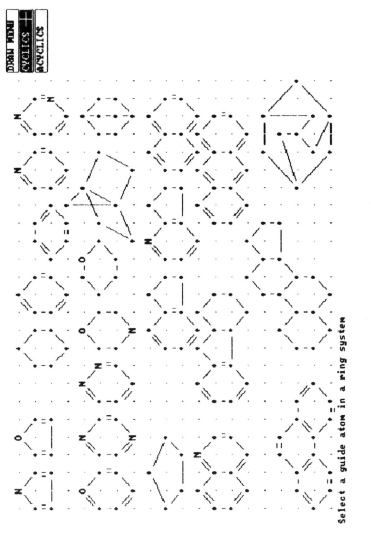

Figure 2.   Cyclic template menu.

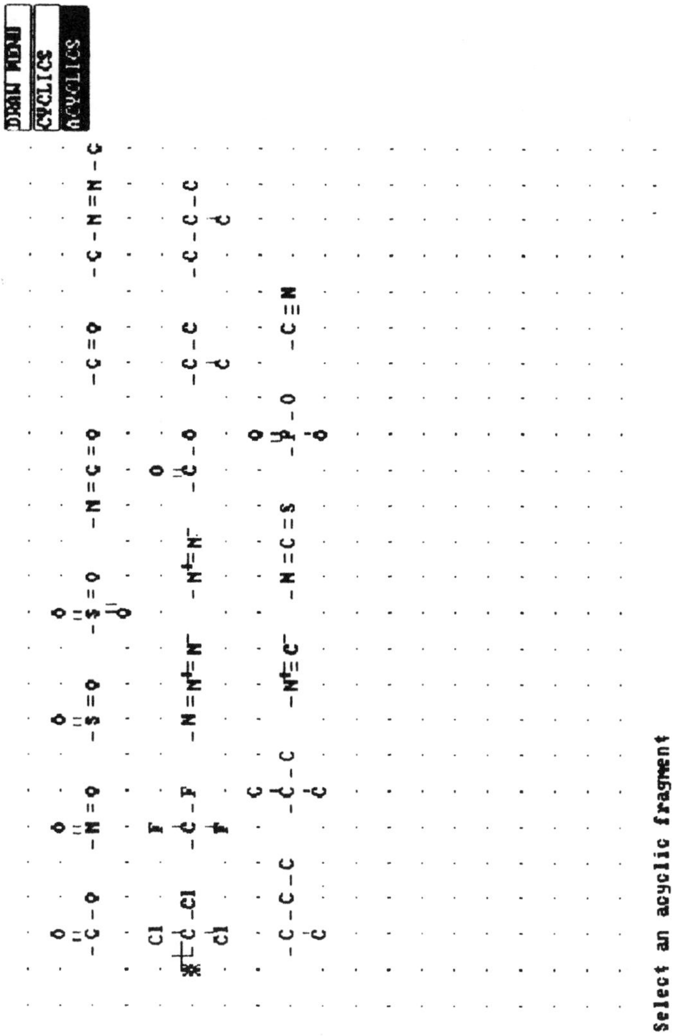

Figure 3. Acyclic template menu.

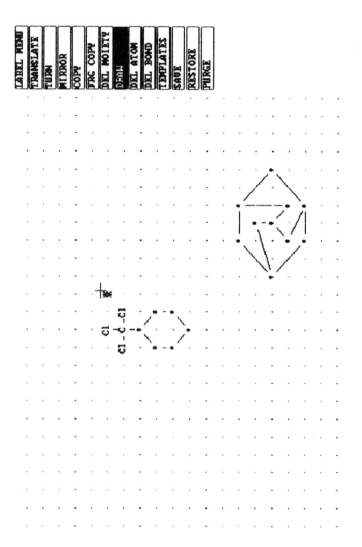

Figure 4.   Phenyl ring with structures from acyclic and cyclic menus.

translation was allowed if atoms would coincide or be off the screen when the operation had been completed.

TURN 90, 180, and 270 would turn a moiety clockwise by the indicated number of degrees. The user would indicate a node about which to perform the rotation and, in the process, select the moiety to be rotated. Since atoms could only appear on the previously mentioned grid of dots, only rotations of 90-degree multiples were allowed. Again, no coincident or off-screen atoms were allowed.

MIRROR X and Y would mirror (invert) a moiety left-to-right or top-to-bottom, respectively. The user indicated an atom to use as the axis of mirroring and selected the moiety be mirrored. As before, no coincident or off-screen atoms were allowed.

COPY copied a moiety and operated exactly like TRANSLATE except the original copy stayed on the screen.

FRAGMENT COPY was the most complex command implemented. It allowed the user to copy a fragment, defined as a moiety attached by only a single bond to another moiety. The user, in this operation, first selected a bond to be "broken" to create a new moiety. Next, the user touched an atom at one end of the "broken" bond to define which moiety was to be the fragment to be copied. Thirdly, the user touched a new node onto which the fragment was to be attached and, finally, indicated where the fragment was to be drawn. This was a very powerful command as it allowed users to perform in one operation what would normally take several: deleting a bond, copying a moiety with possible rotations or mirroring, redrawing the original bond to reconnect the original moiety, and drawing the new bond to connect the moiety copy.

DELETE MOIETY would delete an entire moiety in a single operation.

SAVE allowed the user to save a structure or moiety on the microcomputer's disk for later reuse. The user selected a moiety by touching an atom or selected the entire structure by touching the bottom of the screen. If a moiety were chosen, a local directory was kept and displayed on the screen for the user to update in this operation.

RESTORE was used to restore saved information from the SAVE command. The user was given access to the saved directory for quick selection or could enter a file name on the keyboard for restoring a complete structure.

KILL was the command to delete a saved moiety.

The system implemented for the National Cancer Institute was completed after approximately one work year of effort. It has been used for the entry of approximately 50,000 structures into their chemical database system to date. The structure editor is also used for query purposes by allowing the users to draw a structure or fragment for searching the NCI chemical database system.

Continued Development Work

After we had completed the system for NCI, the software was redesigned to run on an IBM Personal Computer which, by 1985, had available higher resolution graphics adapters, with the Hercules Graphics Board (Hercules Computer Technology, 2550 Ninth Street., Berkeley, CA 94710) being one of the most popular. The Hercules board supported 720 horizontal by 348 vertical pixel resolution and

was close to the computed requirements needed to edit large
structures adequately. In addition, since the reimplemented system
would be used in large part for only query purposes, where the
structures or fragments generated would be much smaller than would
be entered into a database, the very low resolution of the standard
Color Graphics Adapter (at 640 by 200 pixels) could be tolerated
even though the esthetics of stair-stepped bonds and crude fonts
were minimally acceptable.

Although the TOUCHPEN was available for the IBM PC, the cost
was considerably higher than that of the mice then available, the
suppliers of mice were well established financially, and user
acceptance of mice was general. Thus, the TOUCHPEN interface was
replaced with one for a mouse.

At this same time, several new graphics interfaces were rapidly
being introduced for the IBM PC and related clones. This raised the
question of whether to attempt to support many of the boards with
privately written interface routines (which was done for the
Hercules and IBM Color Graphics Adapter boards) or to look for a
more general approach. Of the general approaches available at the
time, only two were feasible and offered by substantial companies.
One of these was the Halo package by Media Cybernetics (Media
Cybernetics, Inc., 7050 Carroll Avenue, Takoma Park, MD) and the
other was Graphic Software Systems Virtual Device Interface (Graphic
Software Systems, Inc., 9590 SW Gemini Drive, Beaverton, OR 97005)
which was also sold by IBM as the Graphic Development Toolkit.

Both packages were designed to support multiple graphics output
devices as well as several input devices (such as the mouse, light
pen, digitizer tablet, etc.). We finally decided on the Graphical
Software System Virtual Device Interface system (GSS VDI) for
reasons of cost as well as support. The Halo package licensing fees
were too high for a program with relatively narrow distribution,
while GSS offered a modest licensing charge with a modest royalty on
each sale, which was much more amenable. Also, since IBM had placed
their imprimatur on the GSS approach by selling the package under
their trademark, it was felt that this system was more likely to
become a standard. A further advantage of the GSS approach was
their support of a wide variety of output devices, including not
only traditional plotters, but inexpensive dot-matrix printers.
This allows one to produce structure hard copy with resolution
limited by the individual device, not by the resolution of the
lowest resolver in the system.

The difficulty with most general approaches to computer
problems is a trade-off between speed and generality. In the case
of the GSS VDI package, the early implementations were almost
unacceptably slow even though they did support many of the graphics
boards being sold. As it turned out, GSS released a new version of
their system, which was much faster and had acceptable speed, while
our development was going on. The current GSS product is called the
Computer Graphic Interface (CGI) and differs from the VDI system
mostly in terms of scope, while maintaining remarkable speed for the
generality of their approach. Unfortunately, there is still no
standard for graphics in the microcomputer world (at least in the
IBM-compatible area), but if one were to choose an implementation
tool for graphics today the options would have to include Microsoft

Windows (Microsoft Corporation, 10700 Northup Way, Box 97200, Bellevue, WA 98009).

Several other changes were made to improve the structure-drawing interface. The menu was trimmed down so that only two screens were necessary for all work by eliminating the separate MOVE menu. This eliminated the delays encountered in switching to the screens used for the MOVE menu. To compensate for the reduced number of menu items initially displayed to the user, sub-menus were displayed along the bottom of the screen for those modes that required them. Now, LABEL ATOM, LABEL BOND, SET MASS, and SET CHARGE are consolidated into a single menu entry which then displays the sub-menu for selection of the four above modes. Finally, a third level of menu would be diplayed for selection of the atom symbol, bond type, charge, or mass. (Mass is entered from the keyboard, with a prompt appearing in the message area.) Similarly, instead of DELETE ATOM, DELETE BOND, and DELETE MOIETY, the main menu has simply DELETE with a sub-menu being displayed to prompt for the type of delete to perform; COPY MOIETY and COPY FRAGMENT were merged into just COPY. The KILL command has been replaced with less bloodthirsty but equally final PURGE.

In the original version delivered to NCI, it was difficult at best to create or edit the cyclic or acyclic menus: no operation was explicitly defined for the user since these menus would never need to be altered for NCI's application. In SuperStructure, editing of these two menus is easily accomplished using the same SuperStructure techniques that one would use for any structure.

With the change from using the TOUCHPEN and further familiarity with user needs it was decided to drop the NORMALIZE option in SuperStructure. The positioning accuracy of the mouse was superior to that of the TOUCHPEN and the better view of the screen (unhampered by the pen, one's hand, and arm) as well as the improvement in responsiveness led to this decision. This also eliminated the need for two cursors, one for the pointing device and one for the atom location that would be selected when in NORMALIZE mode.

The operation of labeling atoms with symbols and charges or masses was changed so that these operations are completely independent rather than linked as they were previously. In SuperStructure, the user may select any of these modes and do the relabeling immediately without having to enter or reenter the symbol, for instance, if it had not changed.

SuperStructure, when started, allows the user to select either interactive or stand-alone modes so that off-line creation of structures for entry into a database or for query can be effected, reducing the costs of communication or searching. It also allows the user manually to select the Tektronix 4010 emulation mode in cases where that mode is not selected automatically by the host computer.

Figure 5 shows the final version of SuperStructure with the consolidated menu scheme and a complex structure (Vitamin B12) to illustrate many of the features available.

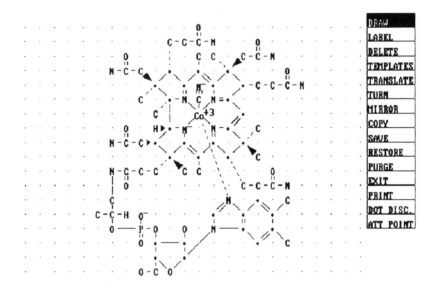

Figure 5.   Final SuperStructure menu with Vitamin B12 structure.

Summary

If development of a graphics structure editor were to be attempted today, many of the same problems would still exist in design and implementation. We chose a path which produced an end result that was very well accepted by the user community and resulted in significant savings in the entry of chemical structures. As a measure of the success of the design and implementation, the program which resulted from the initial implementation for NCI has been modified and improved and is now being incorporated in the Derwent Chemical Patent Database system as well as being the only graphics input query mode for The Chemical Information System.

Literature Cited

1. Lemmons, Phil. BYTE November 1982, 216–254.

RECEIVED March 11, 1986

Chapter 8

Structure Graphics In: Pointers to Beilstein Out

Alexander J. Lawson

Beilstein-Institut, Varrentrappstr. 40–42, D–6000 Frankfurt 90,
Federal Republic of Germany

The automatic correlation of organic structure
graphs with the corresponding reference position in
the printed Beilstein Handbook has been implemented
on a personal computer basis and is briefly
described. Further applications of the algorithm
are discussed.

This paper will concentrate on one aspect of the fact that the
Beilstein database is an ordered collection of critically evaluated
data on organic compounds, namely the question of ordering or
sorting chemical structures. Sorting according to structural
criteria implies a certain degree of generic coding in the widest
sense. This is very much a relevant topic in general, as evidenced
by the great interest currently shown in Markush representation ;
the author will make no attempt here to describe other alternative
approaches currently in development or implemented on mainframe
cumputers. However, one must in all fairness point out that the
approach presented here is largely based on principles laid down a
long time ago, (1) by very talented and imaginative researchers at
Beilstein. The new aspect of this paper lies in the realization of
these principles on an automatic basis, at a local microcomputer
level, opening up the way for similarity searches and browsing in
chemical structures in either the printed Beilstein Handbook or in
a larger set of machine-readable structures at a low level of
hardware power.
    The best way to illustrate the principles involved in the
automatic generation of pointers to Beilstein is to consider the
operation of the program SANDRA:

SANDRA is an acronym for :

Structure AND Reference Analyzer

As the name says, SANDRA analyzes chemical structures and suggests
relevant references to the Beilstein Handbook. The program runs on

IBM-PC's and compatibles. One point should be made clear at the outset: SANDRA should not be confused with BEILSTEIN ONLINE . The difference is : BEILSTEIN ONLINE will supply chemical information directly, whereas SANDRA gives no chemical information, but rather outputs <u>where</u> to look in the printed Handbook for the information sought (i.e., yields pointers to the database following a set algorithm).

Since it is central to the present discussion, a short summary of the organization of the printed Beilstein Handbook now follows.

Organization of the Printed Handbook

<u>The Beilstein Entry and the Beilstein Supplementary Series</u>. In the production of the Handbook, all relevant critically evaluated information about any one structure is brought together in one unit, the Beilstein entry. Then the information collected for the next structure is critically evaluated and summarized in the next separate entry. This process is carried out for all the information published in the original literature over a specific time period, and published as a <u>Supplementary Series</u> to the Handbook.

The Series of the Beilstein Handbook and the corresponding time periods of primary literature covered are as follows :

Years Covered	Series
1830 – 1909	H
1910 – 1919	E I
1920 – 1929	E II
1930 – 1949	E III
1950 – 1959	E IV
1960 – 1979	E V

The Main Series (H for the German "Hauptwerk") covered all the organic structures published up to and including 1909 and the succeeding Supplementary Series (E for the German "Ergaenzungs-werk") cover the updates and new compounds. At present the Series E V, which is in English, is being published, covering the literature 1960 through 1979.

<u>The Structure "Axis" of the "Beilstein Grid"</u>. The Beilstein Handbook can be considered as a grid with a time axis (the six Series) and a structure axis. The individual entries containing the information on two (hypothetical) compounds are shown schematically in this way in Figure 1.

In one case it is supposed that relevant information was published fairly continuously up to the present day. In the case of the second compound entries were published in E I and E IV (i.e., between 1910 and 1919, and 1950 to 1959). Now let us examine the question of retrieving information on either of these compounds, simply on the basis of structure alone. The individual entries for any particular compound in the different Series are always back-referenced, so the question reduces to understanding the nature of the structure-sorting used in Beilstein (the so-called Beilstein

System) and then, starting with the newest Series, working back in
time till the relevant information is located.
     One   user-problem undoubtedly lies in understanding   the
Beilstein System. The central question is :

"How do I determine the position of my compound on the structure
axis (i.e., how do I manage the transition from structure to
pointer) ?"

Let us examine this question more closely.

The Beilstein System and Pointers to the Handbook.

Consider the structure shown in Figure 2. The operation of the
Beilstein System on this structure is unambiguous, as it is with
any single complete structure. This structure alone fully localizes
the position on the X-axis of the graph of Figure 1, and the
printed Handbook indicates this position with a series of five
pointers, completely independent of the Series concerned.

For this example, these pointers are :

                    Volume number (13)

                    System number (1823)

                    H-page (348)

                    Degree of unsaturation of registry compound (2n)

                    Carbon number of registry compound (C6)

(where the H-page is that page of the Main Series (H) on which the
compound would be logically localized according to the principles
of the Beilstein System, and the term "registry compound" can be
equated with "leading fragment" in the sense discussed below; see
also reference (2)).

     Figure 3 demonstrates schematically where these pointers can
always be found in practice in the printed work : at the top of
every right- and left-hand page in Beilstein. To repeat: any
organic structure, when operated on by the Beilstein System,
defines which volume to take in your hand, and which System Number
and H-page pointers to look for in the page headers on the righthand side, and which pointers of finer resolution (degree of
unsaturation, carbon-number) to look for in the page-headers on the
left-hand side. In principle, any chemist can learn the rules of
the System and apply them well enough to use Beilstein without
having to use nomenclature as a search criterion. Generations of
chemists have done exactly this, and the program SANDRA merely
automates, simplifies, accelerates and refines this approach. The
user inputs the graphics, and the program outputs precisely the
above-mentioned pointers. The response time for analysis is about 3
seconds on average on an IBM PC AT or comparable microcomputer.

Figure 4 shows the screen output from SANDRA after the structure has been input by mouse graphics (the graphic interface is described by Ditschke, Lentz and Jochum in a separate chapter in this book).

The SANDRA Procedure

General Description.   The separate parts of the process are as follows. Firstly, SANDRA accepts connection table input, in this case from mouse graphics.

Secondly, the Beilstein System algorithm comes automatically into play. This is really a fragment-screening with a heavy chemical bias (i.e., in contrast to most structure-screening algorithms, the fragments are classified inter alia according to chemical function). In this way, each input structure is automatically and reproducibly reduced to a set of fragments.

Thirdly, the results of the screening for each fragment are converted to a 12-digit code with a high structural resolution, but with no regard for statistical frequency of occurrence in the primary literature. For example, the code distinguishes between heterocyclic rings with 7 ring nitrogen and 13 ring oxygen atoms, and 7 ring nitrogens and 14 ring oxygen atoms, but gives these types of structure as much statistical weight as phenol, or aniline.

The 12-digit code is so constructed that it faithfully reflects the principles used in structure-sorting in Beilstein.

Fourthly, a look-up table is consulted, whereby the generated 12-digit code is localized somewhere between two postings, and the corresponding pointers for the leading fragment are then output by SANDRA as a range. This point in the table can also be made the basis for a normalized hashcode for each fragment.

As already mentioned, the graphical input is discussed elsewhere in this book; the second, third and fourth steps are now discussed in turn.

The Beilstein System Algorithm. The Beilstein System algorithm splits any given molecule into a set of fragments; the following notes are intended to indicate the broad approach, whereby a fuller description can be found in reference (2).

Fragmentation : the splitting points are C-X bonds, whereby the atom X is a heteroatom and does not belong to a common ring with the carbon atom (C) in question. Splitting is obligatory in all cases as defined here, and furthermore in the case of cyclo/oxo tautomerism (ring-opening of sugars) and cyclic di-esters of diols with dibasic inorganic acids. No fragment can be subdivided into smaller fragments by the breaking of a C-C linkage, which means that atoms which are common to more than one fragment can only be heteroatoms.

Classification of features of the fragments : fragments are then classified and coded by taking into account skeletal features, type and multiplicity of chemical functional groups (including masked groups), degree of unsaturation, and carbon number. The

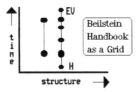

Figure 1.   Entries for two hypothetical compounds
            using the grid model

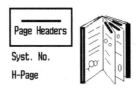

Figure 2.   Structure example used in this paper

Figure 3.   Location of pointers on the Handbook pages

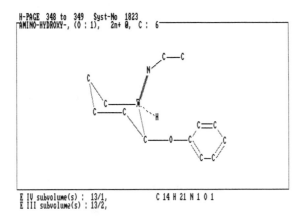

Figure 4.   SANDRA screen for the example used

relative spatial relationship of structural features is not taken
into account. This latter restriction is often a distinct advantage
in similarity searches (e.g. all X-methyl-indole-Y-carboxylic acids
share a common fragment code; the X-ethyl-indole-Y-carboxylic acids
have a code differing only in the carbon number). Much of the
SANDRA code is fairly standard ring-recognition and filters, and
requires no further discussion here. One point of potential
difficulty is nevertheless of interest. The classification of
structural features according to both type and multiplicity
requires a solution to the difficult problem of functional group
recognition in the case of masked groups : the Beilstein
"hydrolysis" scheme is based on a high degree of instinctive
chemical classification as perceived by an organic chemist, and
hence does not appear at first sight to be readily adaptable to an
automatic computer-based analysis.

For instance (and these are relatively simple examples) :
    - chlorobenzene and carbon tetrachloride are regarded as
substituted hydrocarbons, whereas benzoyl chloride is a derivative
of benzoic acid, not benzaldehyde ;
    -bisulfite addition compounds must be recognized as masked
carbonyls ;
    -thiocyanates (R-SCN) must be classified as being ultimately
derived from the parent carbonic acid, while the isomeric
isothiocyanates (R-NCS) belong to the amine class.

    In point of fact, the solution to this problem is remarkably
easy, and lies in the definition that carbons with more than one
(non-ring) heteroatom attached are always regarded as being derived
from carbonyl groups, provided that at least one of the heteroatoms
is other than the attachment atom of a substituent (halogen, nitro,
nitroso, azide). With this problem mastered, the generation of the
12-digit code defines the leading fragment, and hence the registry
compound, and hence the pointers to the database.

The 12-digit Code.   To illustrate this procedure, let us look at
the example used above in Fig. 2 : the structure is input, and the
operation of the Beilstein System on the connection table is then
carried out. The structure is recognised as being composed of three
fragments (substructures) , namely :

    ethylamine   (000500010002)

    phenol   (800100010906)

    and a non-localized amino-cyclohexanol (800510010306)

The 12-digit codes automatically generated for each of these three
substructures are shown in brackets. The order of the numbers or
codes corresponds exactly to the order of the substructures on the
X-axis of the Beilstein System (i.e., to the placing in the
Beilstein Handbook). At this point the largest number is recognized
as being the registry position for the whole molecule i.e., the
compound is registered as a derivative of an amino-cyclohexanol.

The Pointer Lists.   These codes are then compared with a look-up
list, where the individual postings have been chosen manually to
correspond to roughly 30 pages of the Beilstein Handbook Fourth
Series on average. The running numbers (n) of the entries from that
list are shown here for the region n = 1842 to 1851   and this
shows how the last piece of theoretical resolution of the 12-digit
code is discarded in favor of brevity and sensible chemical
grouping. Inspection of Figure 4 and the entry 1849 of Table I
illustrates how the pointers were automatically generated for the
given example.

Table I. Example of the Content of the Pointer Lists

(n)	12-digit code	Syst.No.	H-page
1842	800500030C0A	1803	304
1843	800500030E0C	1804	306
1844	800500031612	1808	314
1845	800500040506	1819	338
1846	800500040D0C	1819	341
1847	800510010304	1823	348
1848	800510010305	1823	348
1849	800510010306	1823	348
1850	800510010307	1823	349
1851	800510010308	1823	349

Generation of a Two-Byte Expression for Each Fragment.   A wider
aspect of the fragment coding should be mentioned here. The value
of this running number (n) retains the inherent sortability of the
12-digit code, but is shorter, and normalizes the content of the
hashcode using actual experience of literature report patterns as
measured by the typeprint of a critically evaluated and
comprehensive collation of data : the Beilstein Handbook. In other
words, a demonstrably well-balanced unit of fragment classification
has been generated on an empirical basis, using a well-tested model
(the Beilstein System). This running number then becomes the basis
for a hashcode for each fragment, and can be stored in 2 bytes.
    On this basis, the structure we have used for our example can
be analyzed in the manner described to produce a series of three
short codes, which describe not only this molecule, but also any
nongeminal combination of the two substituents around the ring (see
Figure 5). This is a really a similarity search in chemical
structures, a capability which is complementary to a substructure
search, and has some relevance for certain types of Markush
representation. A brief inspection of the "chemical environment" of
any randomly chosen region of ca. 30 pages of the Handbook will
demonstrate the strengths and limitations of this approach.

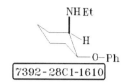

Figure 5. Hash description of the example used

Conclusion

Since each of these fragment codes can be used (separately or in combination) for the inverted lists of a machine-readable set of structures, in principle we can envisage the extension of the browsing effect (for which the Handbook is famous) into a structure file such as that compiled at BEILSTEIN, or other compatible files, large or small. Preliminary tests have shown that the response times are of the order of a few seconds (using a stand-alone IBM PC AT with 30 Mbyte Winchester) for a file of 25000 structures. Clearly, personal computers should not be expected to provide the power of a mainframe in this field, but equally the opportunity clearly exists to exploit the flexibility and comfort of local intelligence. This is really the only justification for creating yet another set of identifying codes for chemical structures ; it is however a powerful one, since it comprises a PC-based conversion of graphics into standard compact codes, which are sortable in accordance with the world's largest sorted set of organic structures (the printed Beilstein Handbook). Further research will show if the use of personal computers for the generation of such fragment screens to enhance the speed and flexibility of structure-graphics interaction with on-line databases has a viable future.

Literature Cited

1. Prager, B.; Jacobson, P. Beilsteins Handbuch der Organischen Chemie (vierte Auflage) ; Vol. 1, p 1.
2. Sunkel, J.; Hoffmann, E.; Luckenbach, R. J. Chem. Ed. 1981, 58, 982

RECEIVED March 11, 1986

Chapter 9

Universal Input Program for Chemical Structures

Clemens Jochum, Christa Ditschke, and Jean-Pierre Lentz

Beilstein-Institut, Varrentrappstr. 40–42, D–6000 Frankfurt 90,
Federal Republic of Germany

The Beilstein Institute is currently developing the
world's largest computerized numerical factual data
base of organic chemical structures. To build the
structure file from the Beilstein Handbook
information and the Beilstein Abstracts, more than 4
million chemical structures have to be entered
manually. For this purpose the microcomputer-based
graphic structure input program MOLMOUSE has been
developed at our institute. The structure, including
stereochemistry, is entered using a mechanical mouse,
the corresponding connection table is calculated and
stored on floppy disks together with the graphical
information and topological stereodescriptors.
Structures can be printed using a laserprinter or a
thermoplotter.

MOLMOUSE (Input of organic MOLecules using a mechanical MOUSE) was
developed within the Beilstein Database Project. The projected,
numerical factual database BEILSTEIN-ONLINE represents a natural
extension of the BEILSTEIN Handbook of Organic Chemistry which has
been published for more than 100 years in more than 300 volumes.

The present BEILSTEIN Information Pool contains Handbook and
Registry data of the literature time frame from 1830 to 1960. Since
the handbook has primarily been published using conventional non-
computerized typesetting methods, this information pool is for the
most part available only in printed form. However, the most recent
25% of the published volumes has been printed using electronic
typesetting methods and is therefore available in a computer-
readable form. (This concerns only the Handbook text. The structural
formulas of the electronically typeset handbooks are not available
in computer readable form.)

0097–6156/87/0341–0088$06.00/0

In addition, the primary literature from 1960 to 1980 has been completely abstracted. Over 7 million paper records contain the structural formula, numerical physical data, reaction pathways and original literature citations (Figure 1).

This information pool will gradually be transferred into a computer-readable form and extended by additional new sources of information (see below). All future Beilstein information products will be generated from this information pool which will be organized in an internal database (Figure 2).

Information Sources of the Database. The Beilstein database will be generated from four sources of information (Figure 3):

1) The printed Handbook-series from the Basic Series up to the fourth Supplementary Series (H to E IV). These series are almost completed and contain the literature from 1830 to 1960. These volumes contain the factual data of more than 1 million organic compounds.

2) The printed Handbook material of the fifth Supplementary Series which contains the literature from 1960 to 1980. As in the previous handbook series, these data have been thoroughly checked for errors and redundancies.

3) The publishing of the Fifth Supplementary Series will probably continue for more than one decade. The primary literature of this time period (1960 - 1980), however, has been completely abstracted in more than 7 million abstract cards.

4) The factual data of the primary literature from 1980 will be extracted in a paperless fashion. The abstracted data are entered directly in a structured manner into microcomputers and stored on magnetic diskettes.

Decentralized Input. The BEILSTEIN Computing Division has developed three microcomputer programs which enable the chemist to enter all the information from its respective source in a structured manner without any specific computer knowledge. All programs work on IBM-PCs and 100%-compatibles with at least 256K of memory and two floppy disk drives:

EKSTASE is used to input data from the BEILSTEIN Handbook. Numeric and factual data are entered separately from the structural formulas as most of the structures are converted automatically into a topological representation (see below). The person who enters the input can choose from various panels to enter different kinds of data (e.g. reactions, physical parameters or citations). To make corrections, he can scroll back and forth on the screen or he can go back to a specific data entry. The systematic IUPAC-names of the reference compounds are automatically copied from the BEILSTEIN Registry tapes so as to avoid manual input errors. These compound names are German for Handbook Series H to E-IV and English starting from E-V (see below).

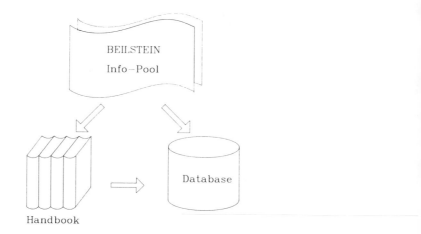

Figure 1. Present State of Progress.

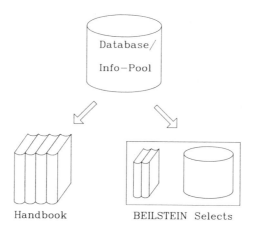

Figure 2. Future Concept.

EXCERP allows the paperless abstracting of the primary literature. The abstracting scientist first enters the structure using a mouse on a PC-compatible graphics screen (see below). After leaving the structure input section, the scientist can choose from more than 25 different screens to enter numerical and factual data in a predefined way according to the BEILSTEIN data structure. Corrections are possible at any time during this input process.

BIZEPS is a stripped-down version of EXCERP for inputting the abstract cards of the Fifth Supplementary Series (see 2.) Since each abstract card usually contains fewer physical parameters than the final Handbook article (which is the summary of many abstract cards), this program is further optimized for input speed by using function keys for entering numerical and factual data. The graphic structure input part is identical to EXCERP.

All three programs store the input information on floppy disk. Most of the input is done by external contributors who send the disks in certain time intervals to our institute for further processing. The data of the different input sources are uploaded into our IBM mainframe computer, converted to the same file format and stored in an intermediate file. After several additional automatic and manual checks the data are loaded under the database management system ADABAS (Figure 4).

To avoid most of the manual input of the Handbook structural formulas, a program has been developed which converts systematic chemical names (IUPAC-nomenclature) to topological representations. On a test data set of approximately 400,000 names, the program was able to convert more than 80% to connection tables without error. It is important to point out that the program either gives a message that it cannot convert a given name with its internal name fragment set or it correctly converts the name. So far, no errors have been encountered. The program currently works only with German names. A version for English names is currently under development and should be finished by the end of 1987.

The Handbook contains more than 1 million structures. Since the name translation program can convert approximately 80% of them automatically into connection tables, there remain 200,000 compounds which have to be entered manually. In addition, the 7 million structures on paper records (see 3 above) have to be entered manually to convert them into connection tables. We estimate that 500,000 additional structures have be entered per year for the primary literature abstracting (1980 onwards (see 4 above)). We therefore had a very high demand for an efficient graphic structure input program, which allows fast error-free entry of structures on a personal computer. The program MOLMOUSE has been developed at our computing division for the last two years to solve this problem.

MOLMOUSE Program Design Principles

The basic design philosophy behind MOLMOUSE was to generate a program which allows the input person (chemist as well as non-

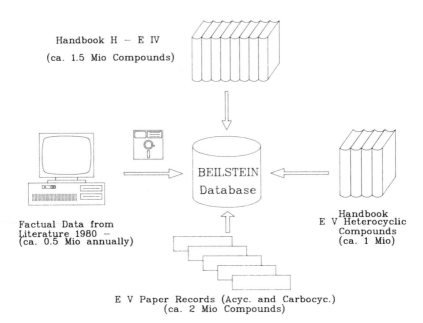

Figure 3. Beilstein Database Sources.

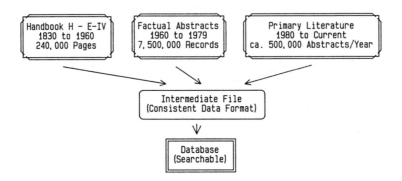

Figure 4. Database Generation.

chemistry-trained input person) to enter the structure in a way similar to that he would use to draw it on paper. The most important advantage of this philosophy is that the entered structure looks very similar to the printed copy from which it was entered. It is therefore very easy to compare the structures for error checking. (Error checking has to be done quite frequently since MOLMOUSE refuses any structure which does not correspond to the manually entered molecular formula.)

Before arriving at this design we had experimented with the development of a keyboard-key-to-fragment assignment program. For an experienced entry person this method also allows a very fast input of chemical structures. The input person, however, has very little influence on the structure's appearance on the screen (or later on an on-line display file). Error checking with this method is rather difficult. It is also much more difficult to learn than a drawing-oriented program. Internal benchmarks also showed that a drawing-oriented program containing a set of callable predefined structures allows a significantly faster input (from chemistry- and non-chemistry-trained persons) than a keyboard-to-fragment program.

We decided on a mechanical mouse as the major input device. The main advantages of this device are:

It works on practically any surface (even on your lap if your desk is too crowded).

It is relatively inexpensive (compared to digitizing pads).

It does not require any special pad to run on (like optical mice).

It has an even higher resolution than most optical mice.

The failure rate is extremely low. We are currently employing more than 150 PCs for structure input and have never had any mouse break-down.

It is much faster and more ergonomic to use than light pens.

The mouse is not the only input device. Benchmarks showed that it is faster to call up certain commands or fragments by pressing a keyboard key rather than fetching the item from a graphic menu by using the mouse. The mouse is only used for drawing bonds and setting atoms (like a pencil). All other commands are single-key commands (easier for non-typists) on the keyboard. Thus the mouse-drawing hand never has to leave the mouse and the other hand can be used to give additional commands on the keyboard.

Two sets of fragments help to increase input speed. The user can choose either a set of 30 program-predefined fragments or he can use his own set of 30 additional user-defined fragments. Both sets of fragments are available at any time during the drawing process. The restriction to 30 fragments for each set is not an intrinsic hardware or software limit, but enables the retrieval of each fragment of a set with one keystroke (function keys F1 to F10 and the CTRL- and ALT-Keys in combination with the function keys). Using the mouse again, the size and the orientation of a fragment can be varied freely on the screen.

MOLMOUSE Input Description. When the input person starts MOLMOUSE he is presented with an almost empty screen for drawing (Figure 5). This screen is only a small part of the whole drawing surface which consists of 64,000 units in x- and 64,000 units in y-direction (64,000 x 64,000 Pixels). The part shown on the screen depends on the hardware and the enlargement factor. This method allows device independent storage of the graphic coordinates of the atoms. On the other hand, since only two-byte integer numbers are used for graphic representation, no floating point hardware is necessary for fast zooming or fragment rotating.

There is a small arrow in the middle of the screen which represents the mouse cursor. The mouse has two buttons. The left button is generally used to set a new atom. The right mouse button is primarily used for correction (changing the position of an atom or removing it).

If the left button is pressed once, the first atom is set. The cursor disappears and a bond is "connected" from this atom to the mouse (similar to a rubber band). The bond order can be changed now at any time using the respective numeric keys 1, 2 or 3 while moving the mouse on the rubber band. The rubber band then changes accordingly to "double bond" or "triple bond". After moving the mouse to the desired location and pressing the left button again, atom 2 is set. As long as the mouse is not moved away from this atom, a small menu appears at the bottom of the screen which allows changing the atomic number, mass (for different isotopes), charge or free electrons (for radicals). (It is also possible to define delocalized charges or radicals). If bond drawing is to be continued on another atom, the left mouse button is pressed again after an atom has been set. The cursor reappears and the placement of atoms and bonds can be continued at another location.

In addition to the different bond orders, there are several other types of bonds to handle stereochemistry (bold, dotted and wavy bonds). Stereocenters are displayed by showing the respective atom in reverse video. For molecules with many stereocenters (e.g. sugar chemistry), the Fischer Projection Method can be used for entering stereocenters. (Figures 6,7)

If an atom and its respective bonds are to be removed from the molecule, the mouse cursor is positioned on top of this atom and the right mouse button is pressed twice. Pressing the right button just once changes the location of the atom and the respective bonds. The atom is fixed at a new location by subsequently pressing the left mouse button. (Left mouse button = constructive, right mouse button = destructive; see philosophy above).

No. of compound:4

Figure 5. MOLMOUSE Entry Screen.

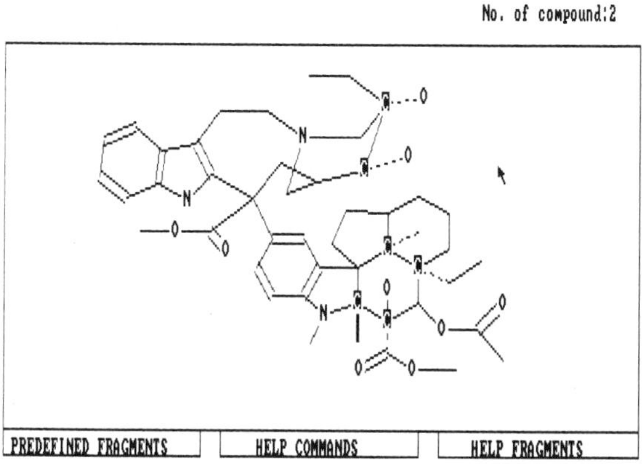

No. of compound:2

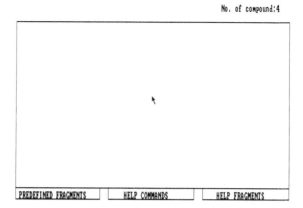

Figure 6. MOLMOUSE Representation of a molecule with several stereo centers.

No. of compound:3

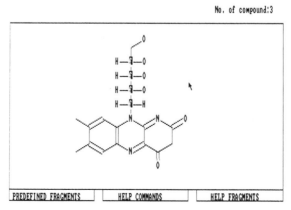

Figure 7. Fisher projection.

Several additional commands allow repositioning of the whole molecule. With the Move command it can be moved to any position of the virtual surface (inside and outside of the screen). Other commands enable the structure to be enlarged or decreased in size.

Pre- or user-defined fragments are called up by pressing the respective function key. The size and the orientation of the fragment on the screen is defined by positioning fragment atoms 1 and 2 with the mouse cursor.

Help for command- or fragment-information is available at any time by positioning the mouse cursor at the respective help fields at the bottom of the screen. (Figures 8,9)

During input, the program always monitors the connectivity of the atoms and warns of "unusual" connectivity for respective atom types. These warnings can be overwritten by adding charges or free electrons to the atom.

MOLMOUSE Program Description.   The MOLMOUSE program can be divided into three parts:

1) The Input and Display Section. This module contains the keyboard single-key command interpreter, mouse- and display- device drivers.

   The mouse movements are handled by an interrupt-driven mouse device driver. Thus keyboard commands can be entered at any time, even during mouse movement.

2) CT-Generator. This part contains the main module for the generation of connection tables and virtual (device-independent) graphic coordinates.

3) The File Handling Section. In this section the connection tables are compressed to a very small binary storage format (see below). This part contains subroutines for storing compounds and reading in predefined fragments from fragment files. (See MOLMOUSE File Format below.)

Besides producing a hardcopy on a graphics matrix printer, a separate utility can generate a high-resolution hardcopy on a laser printer (Figure 10).

The first program version which is still in use for Beilstein input production was written in compiled Microsoft Basic. So far we have input more than 500,000 structures using this program. Currently, more than 100,000 additional structures are input each month.

As described above, this program not only exists in a stand-alone version, but is also part of the EXCERP and BIZEPS factual data entry programs. In addition, the input and display section of

```
A    Alternate Bond Order
B    Backup User-defined Fragment
C    Center Structure
D    Decrease Size of Structure on Screen
E    Enlarge Size of Structure on Screen
F    Fischer cross
G    Get any Structure from Disk
K    Kill Structure without Backup
M    Move Structure
N    Number the Atoms
P    Paint Screen after Erasing
Q    Quit (end of structure input)
R    Radicals and Charges (Delocalized)
S    Symbols on Atoms
U    Stereo Descriptors
W    Without C-Symbols or Numbers
X    Bold Line
Y    Dotted Line
Z    Wavy Line
1    Single Bond (Default)
2    Double Bond
3    Triple Bond

                    Press any Key to Continue
```

Figure 8. Command Help Screen.

F1	6-Ring (unsat.)	F2	n-Ring (unsat.)
F3	6-Ring (sat.)	F4	n-Ring (sat.)
F5	C-COOH	F6	C-CN
F7	C-SO3H	F8	O-SO3H
F9	O-SiMe3	F10	O-PO3H2
CTRL-F1	C-nitro	CTRL-F2	C-diazo
CTRL-F3	C-azido	CTRL-F4	benzoyl
CTRL-F5	benzyl	CTRL-F6	trityl
CTRL-F7	t-butoxycarbonyl (Boc)	CTRL-F8	benzyloxycarbonyl (Z,Bzo,Cbo)
CTRL-F9	toluol-4-sulfonyl (tosyl)	CTRL-F10	trifluoracetyl
ALT-F1	Gly	ALT-F2	L-Ala
ALT-F3	L-Val	ALT-F4	L-Leu
ALT-F5	L-Asp	ALT-F6	L-Glu
ALT-F7	L-Pro	ALT-F8	L-Phe
ALT-F9	L-His	ALT-F10	L-Try

Press any Key to Continue

Figure 9. Predefined Fragment Help Screen.

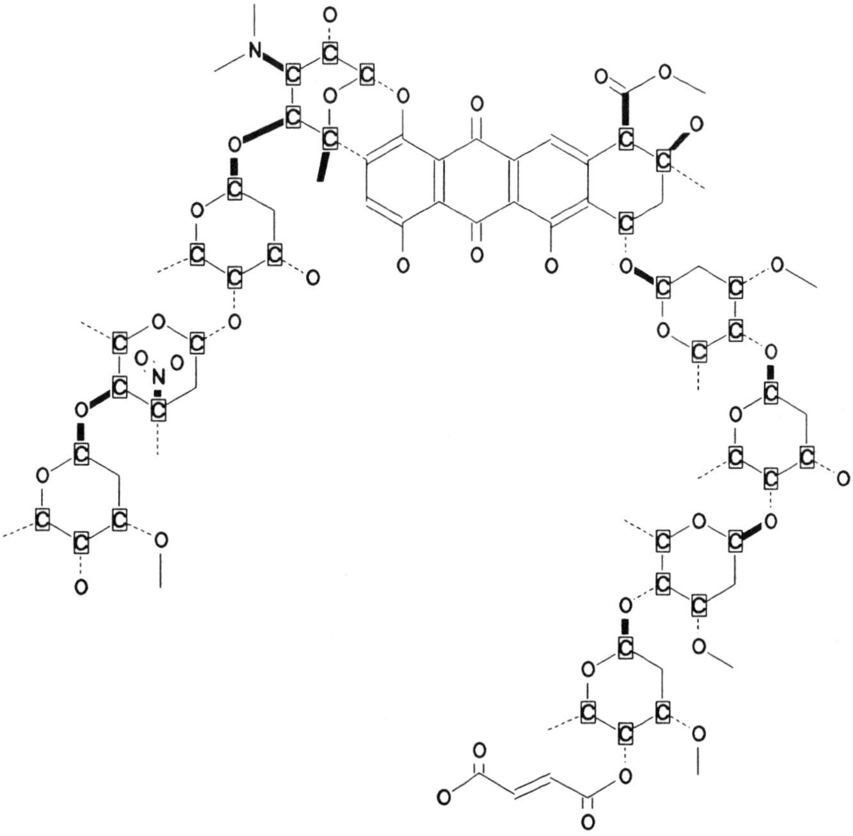

Figure 10. Sample Laser Printouts.

MOLMOUSE is used within SANDRA to allow easy retrieval of Beilstein Handbook compounds. (The SANDRA program is described in a separate chapter of this book.)

We are currently testing version 2 of MOLMOUSE. This version has been completely rewritten into the C language. It will support more display devices than Version 1 and will be implemented in other Beilstein programs currently under development (see below).

All constitutional and stereochemical information is stored topologically in a structure file:

The first record defines the number of stored compounds as well as additional administrative information.

The first record of each structure contains the number of atoms, Fischer projection centers, charges, radicals, isotopes etc. and additional administrative information.

The logically subsequent records contain the connection table in compressed format including all stereochemical bonds and the virtual (device-independent) graphic coordinates for each atom.

The length of the file is only limited by the size of the storage medium.

Hardware Requirements. MOLMOUSE will run on any IBM-PC or AT or 100%-Compatibles which are equipped with at least 256K Bytes RAM. Most of Beilstein's 150 external contributors use the program on an Olivetti M-24 (this PC is called AT&T 6300 in the U.S.) or a Compaq Deskpro equipped with two floppy disks.

Version 1 supports only the IBM color graphics adapter (CGA) and compatibles (e.g. Compaq, Olivetti, AT&T, Toshiba, Paradise Display adapter, Genoa Spectrum Display Adapter etc.) as a display device. This also includes support of the IBM enhanced graphics adapter (EGA) in its CGA emulation mode.

Version 2 will in addition support EGA in high resolution mode and Hercules graphics.

Since the graphics coordinates are stored in a device independent format, structure files entered on one device can also be displayed or modified on another display device.

Future Perspectives

MOLMOUSE will be used as an input device for the Beilstein Structure/Substructure Search System currently under development. Both the PC- and the mainframe-based version of this system will use MOLMOUSE for input.

We are also planning a "CAD" version of MOLMOUSE for electronic typesetting of the Beilstein Handbook. This version will work on a

graphic   workstation   (e.g.   IBM 6150) with a very   high   resolution
screen.   This was one of the essential reasons for rewriting Version
2 into C.

Using   another program developed at the Technical University in
Munich,   Cahn-Ingold-Prelog-like   stereocenters are calculated   from
the   stereochemical   bonds.   This program   together   with   automatic
tautomer and canonization routines convert the MOLMOUSE input format
into the Beilstein Database File Format.

RECEIVED March 17, 1986

Chapter 10

A Data Base System That Relies Heavily on Graphics

G. W. A. Milne

Information Technology Branch, Division of Cancer Treatment,
National Cancer Institute, Bethesda, MD 20892

The National Cancer Institute operates several large,
numeric databases and is taking increasing advantage of
graphics as a means of presenting large amounts of data.
The technology and philosophy underlying such work is
described in this paper.

Since 1955, the National Cancer Institute has supported a program
in which large numbers of chemicals are tested in an attempt to
identify compounds which possess activity against human cancer.
The program has had some success in that of the approximately 40
anti-cancer drugs that are currently commercially available in the
U.S., almost all were discovered by or developed in this program.
These 40 drugs emerged however from a starting group of about half
a million compounds, and it is now clear that a relatively large
number of compounds must be examined in order to find one useful
agent.

Large databases are therefore to be expected in this program
and in fact, the NCI Drug Information System (DIS), which carries
all the data associated with this effort, is currently storing
about 4 billion bytes of data. With such large files, even
legitimate and correct queries can often produce prodigious amounts
of information and accordingly, a major task for the DIS has been
to design methods for presentation of data that provide concise
reports. Graphics are very valuable in this connection and the use
of graphics in the DIS forms the subject of this paper.

The NCI Drug Information System

The operation of the drug screening program and the DIS is illustrated in Figure 1. The acquisition step (#2 in Figure 1)
represents the first DIS operation for a compound. From a variety
of sources, including literature surveillance and liaison with
industry and academia, the program identifies each year some 50,000
structures judged to be potentially of interest in connection with
cancer chemotherapy.

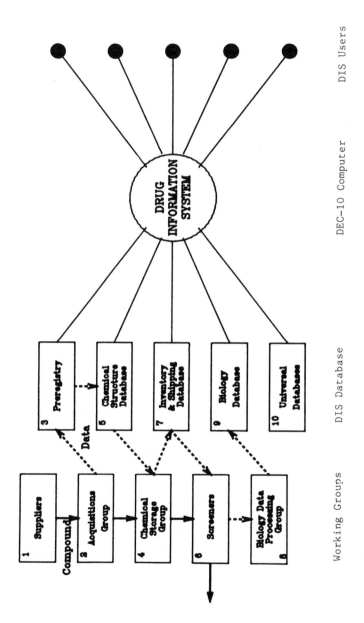

Figure 1. Overall Schematic of the DIS.

These structures are all entered into the Preregistry database of the DIS. Using criteria such as structural uniqueness and estimated probability of activity, computer programs identify about 20,000 of the better candidates from this input and the NCI requests samples of these compounds for testing.

Approximately half the compounds requested are received. As they are acquired, they are assigned a permanent "NSC Number" and their chemistry records are moved from the Preregistry to the Chemistry database. The physical samples are labeled with their NSC Numbers, in barcoded form, and transferred to a storage facility, where they are logged in. This second contractor weighs the material and creates an Inventory record for that sample. A Shipping record is also begun at this point, reflecting the fact that the compound was shipped on given dates from the Supplier to the Acquisitions contractor and from the Acquisitions to the Storage contractor. These new records are used, respectively, to update the Inventory database and the Shipping History database.

For preliminary testing, which is against P388 leukemia in mice, the DIS controls the flow of compounds from the Storage contractor to the various screening laboratories. As a screener's load/capacity ratio drops, the DIS automatically directs more compounds to be sent to that screener. The capacity of a screener can be adjusted by NCI staff so as to reflect the screener's contractual obligation. The storage contractor receives such shipping requests from the DIS and fills them on a daily basis. Each year, there are some 10,000 such "automatic shipments", and in addition, some 2,000 individual shipments of compounds destined for secondary testing are ordered by NCI staff.

The screening laboratories use a full-screen edit program operating on a Hewlett-Packard HP-2645A terminal to collect the data from completed screening experiments (#6 in Figure 1). Once all the data have been entered, they are written in condensed form onto a tape cassette in the terminal. At regular intervals, typically daily, the terminal is logged onto the NIH computer facility and the contents of the tape are downloaded into the NIH IBM 370 computers. There, the downloaded files are used as input to a program (#8 in Figure 1) which examines all new data for internal consistency and freedom from logical errors, and then calculates the final test results from the raw data. Errors that can be corrected on the spot are resolved; other errors that are detected are passed back to the screener. When they logon next, the calculated data and the errors are presented for resolution before more data entry begins. When data have been finally validated in this way, they are written to a staging area to await the next master file update. These updates are carried out every two weeks and trigger an update of the online searchable files in the DIS. Such a biweekly update is reflected in the content of the searchable biology database (#9 in Figure 1).

DIS Databases

There are 24 linked databases in the DIS. Many of these are quite small, and a few are not directly accessible to users. The major files are all interactively searchable, and these are shown in Figure 2. Each of the databases contains some number of fields, and each field is identified by means of a "field mnemonic", which is usually a four-letter code, such as ADDR for address, or MOLF for molecular formula. There are 360 distinct fields in the DIS; 232 of these are searchable and all of them can be displayed on command. Every one of these field mnemonics is unique; it is therefore unnecessary for a user to remember which database is being addressed, because the DIS can recognize the field mnemonic and search the appropriate database.

DIS Computers

The DIS runs on computers of the NIH Computer Center which are shown schematically in Figure 3. Most of the DIS code and data is resident upon a DEC System 10 computer (#2 in Figure 3). This is linked to an IBM 3091 system (#3) which, in turn has a Hewlett-Packard 2680 high-speed (0.7 secs/page) laser printer (#5) along with its controller, an HP-3000 minicomputer (#4), configured as a peripheral device. This design is somewhat complicated but is mandated by various constraints that are beyond NCI's control. Delivery of graphics from the DEC-10 to the laser printer is generally handled by macros that are built into the DIS and the operation is sufficiently smooth that the printer can be regarded as though it were a DEC-10 peripheral. Laser printers of this sort have full graphics capability with moderately high resolution and it is on this printer that all the graphics from the DIS are printed. Other output devices are accessible from the DIS and these include Calcomp and Zeta plotters, as well as an Apollo workstation, which is used to support molecular modeling.

Graphics from the DIS

This Section contains a description of three distinct areas where the DIS makes extensive use of the graphics printing capability that has been described.

Printing of Letters. Each year, NCI generates as many as 20,000 letters to suppliers of compounds. Such a volume of correspondence must be generated automatically, and programs to handle this have been installed in the DIS. The quality and style of the letters is important because NCI maintains a collegial relationship with its suppliers. The letters therefore are personalized to some extent, routinely cite prior correspondence, carry structure diagrams and are often written in languages other than English. The overall function of the letter generating program is shown in Figure 4. Once a decision has been made to try to acquire a compound, it is determined whether this is a first order for the compound or a refill. In either case, the prior records for the material are reviewed and data that are to be cited in the new letter are retrieved. Then the language for the letter is selected. Where appropriate, the DIS uses French, German, Spanish or Japanese. Otherwise, English is used. Once the language is chosen, a font selection must be made, the correct "canned" text retrieved, name,

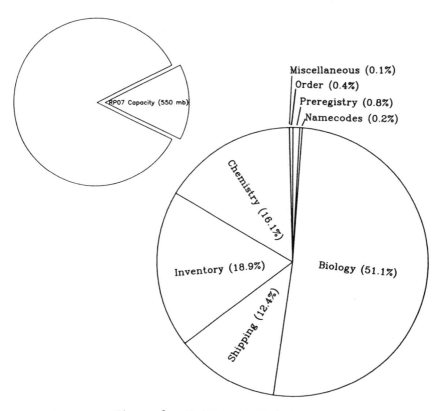

Figure 2.   Major DIS Databases.

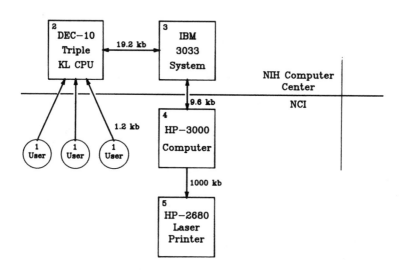

Figure 3.   Computers Used by the DIS.

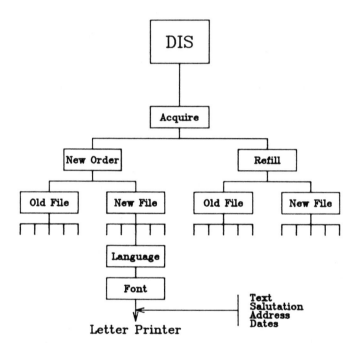

Figure 4. The DIS Letter Generator.

address, and salutation added and dates that are to be cited in the
text must be inserted. Then the letter is sent to the printer,
usually as one of a large batch of letters. The printer prints a
letterhead on the first page, then prints the date and the name and
address of the recipient. It then switches language and font, as
necessary, and prints the letter and finishes the first page by
printing a signature block using the normal pica font. A signature
- which is just another graphic, but one that is stored under
password control - may or may not be included in the signature
block. The second page is an "attachment" which carries details of
a chemical. No letterhead is printed, but the chemical structure
and other identifying data are printed. Here a pica font is used,
irrespective of what was used on the first page. Thus the letters
with their respective attachment sheets are printed in order and
can be mailed directly. The recipient's address is always printed
in English and positioned so that a window envelope may be used for
mailing. Such letters are produced routinely by the DIS at the
rate of a few hundred per week. A letter and attachment to a
Spanish supplier is shown in Figure 5 and Figure 6 shows a letter
to a Japanese investigator.

Chemical Structures. The DIS is required to print hundreds of
chemical structure diagrams each week. These diagrams are used in
all manner of reports and a basic requirement of the system is that
the diagrams be of a high graphical quality. To meet this demand,
the DIS proceeds as follows.

      Structures are entered into the database at the time the
chemical is being considered for acquisition (#2 in Figure 1).
Structure entry is carried out using a microcomputer which is
on-line to the DEC-10. The flow of data during this process is
shown in Figure 7. A program on the microcomputer supports entry
of the structure as a vector diagram which can be modified by the
user until it is chemically and esthetically satisfactory. Then
the vector set is uploaded to the host where it is transformed to a
standard connection table. The connection table is used to perform
a number of checking functions. Once the structure has been
accepted as correct, the connection table is passed to numerous DIS
programs which use it to generate search keys, structure diagrams
which can be typed on a non-graphics terminal and diagrams which
can be drawn on a CRT. Meanwhile the vector set is passed forward
unchanged and is stored as a part of the compound's permanent
record. It is used whenever the structure of the compound is
subsequently designated for printing. These three different types
of structure output are shown in Figure 8, from which it might be
noted that the quality of the diagrams is roughly proportional to
the cost of the output device that is used. The vector diagram
printed by a laser printer is clearly the optimum for high quality
structure diagrams and it is used throughout the DIS when structure
output is to be printed.

Representation of Biological Data. Biological data is recorded by
the DIS in minute detail. Even a simple preliminary test with a
compound in cancer-bearing mice leads to several hundred lines of
data and by the time an active compound comes into consideration

DEPARTMENT OF HEALTH & HUMAN SERVICES   **Public Health Service**

National Institutes of Health
National Cancer Institute
Bethesda, Maryland 20892
TLX:908111

20 August, 1986.

Dr. Benjamin Rodriguez
Chief. Natural Products Dept.
Institute of Organic Chemistry
CSIC
Juan de la Cierva, 3
Madrid 6, Spain

Estimado Dr. Rodriguez:

Incluimos el archivo químico preparado para uno
de sus activos potenciales contra el cancer sometidas
07/29/86. Este archivo contiene la información química
y el número NSC correspondiente para sus muestras.

Este número NSC les identificará su compuesto en toda
futura correspondencia y aparecerá en los datos de
selección. Reportes de los datos de selección se
enviarán a ustedes tan pronto sean disponibles.

Gracias por su interés y participación en el programa
de Quimoterapia del Cancer. Si tuvieran alguna pregunta
ó comentarios sobre esto en el futuro sírvase comunicarse
con nosotros.

                              Sinceramente,

                              Matthew Suffness, Ph.D., Chief
                              Natural Products Branch
                              National Cancer Institute
                              Landow Building, Room 5C-09
                              Bethesda, Maryland, 20892
                              U.S.A.

Figure 5.   Letter to a Spanish supplier.   Continued on next page.

NSC 610854

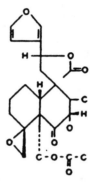

DRUG SYNTHESIS AND CHEMISTRY BRANCH, DTP, DCT, NCI
NSC NUMBER LIST

Dr. Benjamin Rodriguez
Chief. Natural Products Dept.
Institute of Organic Chemistry
CSIC
Juan de la Cierva, 3
Madrid 6, Spain

Transmittal Letter or Collection Date: 07/29/86
Supplier Compound ID: Capitatin, 7-Deacetyl
NSC Number: 610854-G
Molecular Formula: C22H26O8

Figure 5.--Continued.

DEPARTMENT OF HEALTH & HUMAN SERVICES   Public Health Service

National Institutes of Health
National Cancer Institute
Bethesda, Maryland 20892
TLX:908111

20 August, 1986.

Dr. Yoshiyasu Shitori
MECT Corporation
Mitsui Bldg. 5F. P.O. Box 212
2-1-1, Nishishinjuku
Shinjuku-ku, Tokyo 160, Japan

Dr. Shitori ハカセトノ

ハイケイ

キデン カ゛ 07/31/86 ニ ソウフ ナサイマシタ ツイカ テイシュツ フ゛ンノ
サンフ゜ル ヲ マチカ゛イ ナク ウケトリ マシタ。 コンコ゛ トモ コノ カコ゛ウフ゛ッ ハ
ワレワレ ノ スクリーニンク゛ テ゛ハ ヒキツツ゛キ サキニ コノ カコ゛ウフ゛ッ ニ アタエラレタ
ト オナシ゛ NSC ハ゛ンコ゛ウ テ゛ シキヘ゛ッ サレマス。

コノ カコ゛ウフ゛ッ ニ カンスル セイフ゛ッ シケン ノ ケッカ ハ イセ゛ン ト オナシ゛ ヨウニ
チョクセツ キテ゛ン ニ オクラレル ヨテイテ゛ス。

ワレワレ ノ イライ ヲ ココロヨク ヒキウケテ クタ゛サイマシテ アリカ゛トウ コ゛サイマシタ。

ケイク

ナラヤナン ハカセ

V. L. Narayanan, Ph.D., Chief
Drug Synthesis & Chemistry Branch
National Cancer Institute
Landow Building, Room 5C-18B
Bethesda, Maryland, 20892
U.S.A.

Figure 6. Letter to a Japanese Supplier.

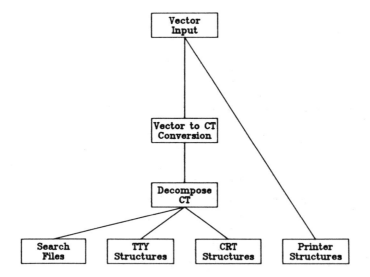

Figure 7.   Entry of Structures into the DIS.

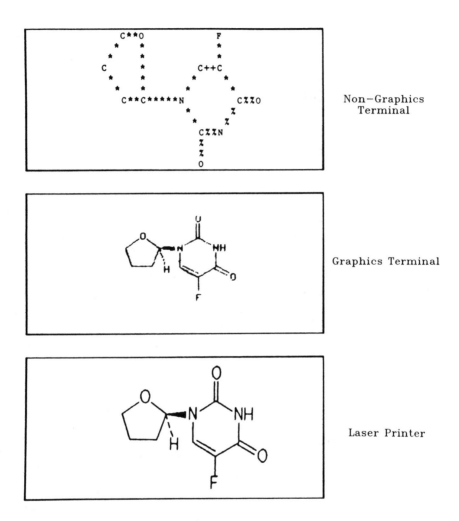

Non−Graphics
Terminal

Graphics Terminal

Laser Printer

Figure 8. Structure Output from the DIS.

for human trials, the accumulated biological data that have been
measured on it typically will require hundreds or even thousands of
pages for a listing. It is almost impossible to assimilate such
quantities of data and many years ago, NCI staff designed a
"Screening Data Summary" which condenses the data greatly, reducing
the page count of an alphanumeric report by about one order of
magnitude. Such a summary may still be 100 pages in length and
difficult to digest and so recourse has been made to graphics
representations of the data to condense it still further. A
bar-graph format was developed for the screening data and is shown
in Figure 9. Each tumor system tested has one or more bars associated with it. The identity of the tumor system is under the
bar and the width of the bar is proportional to the number of tests
carried out in that system. The height of the bar gives the
highest observed activity, or %T/C. This is measured with reference to the left vertical axis (for survival, or life-span systems)
and the right vertical axis in solid tumor regression systems. The
arrow under the bar serves to remind one as to which axis is
applicable. The center of the "X" in the bar represents the mean
of all the %T/C values obtained and the vertical height of the X
provides the standard deviation in the data. Other legends in this
diagram include the device (square or triangle) at the foot of the
bars. This indicates the drug administration route; the square
means intraperitoneal, the triangle, intravenous. The black
horizontal bars show the level of activity that is the pass/fail
criterion currently in use at NCI. The smaller bar graph at the
bottom of the Figure provides dose level information. Each single
dose is drawn here, referenced to the ordinate labelled "DOSE AMT"
and the abscissa calibrated in days. Thus in the first case,
Q01DX09 is standard medical terminology for "once per day on days 1
through 9", and there are therefore 9 vertical bars, one for each
day beginning on day 1; the height of each bar corresponds to a
dose amount of 150 mg/kg of mouse body weight.

A more adventurous variation on the bar-graph format is shown
in Figure 10. Here, instead of a bar, we use a flower to carry the
data. The height of a flower's stem indicates the median %T/C
found with that drug against the tumor, whose identity can be found
at the foot of the flower. The stem height is referenced to the
vertical axis towards which the flower is leaning. The vertical
measure of the bud, or center, of the flower shows the standard
deviation in the data while the number of petals possessed by the
flower conveys the number of completed experiments. The horizontal
width of the bud is inversely related to the dose amount used and
finally, the triangle drawn on the stem of a flower, if present,
says that drug administration in this case was intravenous. Both
the bar graph and the flower garden provide for significant compression of data. Just as the screening data summary was about 10%
of the size of the raw data dump, so these graphics both are about
10% of the screening data summary, in terms of page count.

A more subtle property of the flower garden, which is not
possessed by the bar graph, is that because readers all can distinguish healthy flowers from others, it is possible to make "successful" tests identify themselves. In Figure 10, the two most robust-

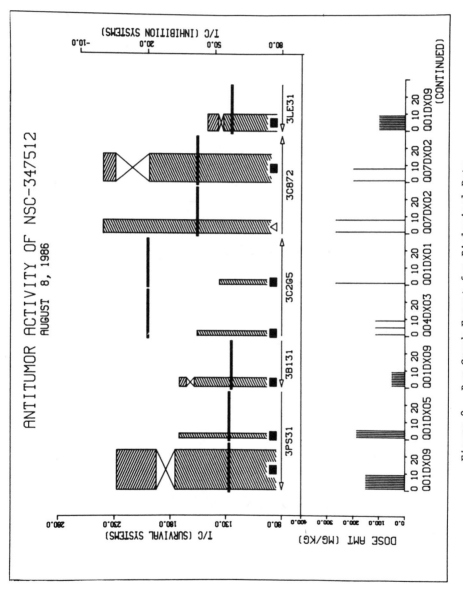

Figure 9. Bar-Graph Format for Biological Data.

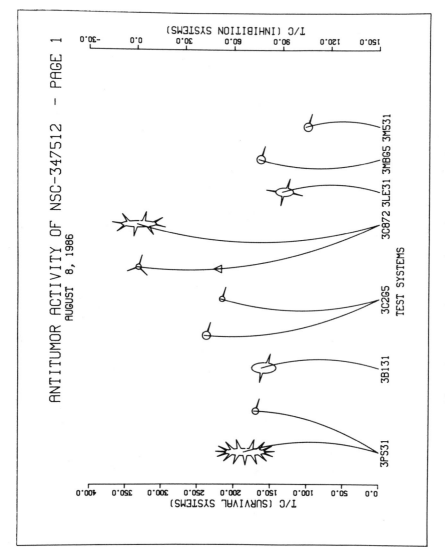

Figure 10. Flower-Garden Display of Biological Data.

looking flowers are the first and the seventh - i.e. those with
many petals and at least modest height. These are in fact the two
most significant tests; It was this compound's activity against
3PS31 (leukemia) that led to its continued testing and the dis-
covery of major activity in 3C872 (colon).   The compound is now
being actively tested against colon cancer in humans. The point of
interest is that someone with little scientific background would
probably have nominated these two flowers as representative of the
"best" tests, and they would have been right.   Not only then is
there much data in this diagram, there is also an implicit "data
key" which makes assimilation of the information easier.

Expert Systems.

A quite different way to deal with the problem of voluminous output
is to condense it by means of an intellectual summary. The Screening Data Summary described above is non-intellectual in that it
merely discards some data and reformats the remainder. A role that
can be played by computer programs, sometimes termed "expert
systems", is to analyze all the data for significant content, using
a set of rules, and then generate a report based upon that significant content.

     In the NCI, one is frequently asked for the status of a
chemical which is somewhere in the multi-year testing cycle. In
such a context, "status" implies a skeletal description of the data
on the compound along with its position in a temporal sense (what
has been done? what remains to be done?). Also loaded into the
idea is some sort of a performance rating. Has it failed trials?
Is it expected to fail trials? Will it make it as far as the
clinic? And so on.

     A program developed within the DIS makes a start on responding
to this sort of query. It simply reviews all the data on a compound and then produces a one-page report like the one shown in
Figure 11. The basic information on the compound, its identity and
its supplier, are given. Then current inventories are reported and
finally, a very concise history of its biological testing is
developed and used to finish out the report. The document that is
produced is very short - always less than one page - and as such is
popular with senior management because it tells what they need to
know without descending into excessive detail. A document of this
sort cannot be produced readily if graphical printing and variable
fonts are not readily accessible and therefore this should be
regarded as another type of graphical output.

Summary

If language or numbers represent artificial means of human communication, then graphic display is a more fundamental mode. It is
at the same time more powerful and easier to understand - an
unusual combination of positives. Computer manipulation of
graphics has always been more difficult than the common alphanumeric computation and this has been an impediment. As technology improves however, generation of graphics has become simpler

NSC-600000

The compound with NSC number 600000 is 1,3-Diazaspiro[4.5]decane-2,4-dione, 3-[4-[bis(2-chloroethyl)amino]butyl]-. Its molecular formula is C16H27Cl2N3O2. The compound's formula weight is 364 and its structure is shown below:

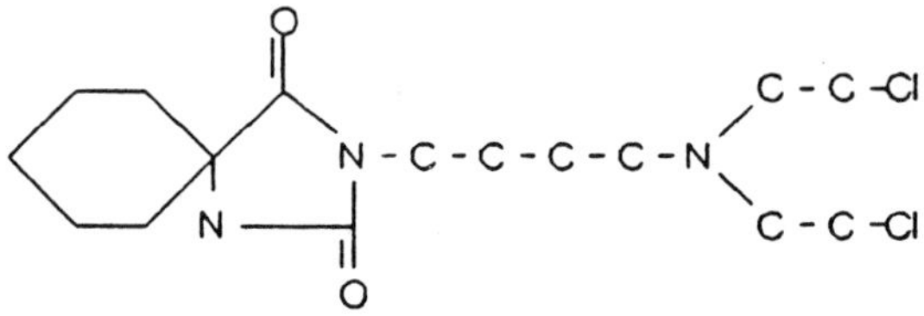

The first sample of NSC-600000 was obtained by NCI on 22-Feb-85 from:

    Dr. John Driscoll
    Drug Design & Chemistry Section
    LMCB, DTP, DCT, NCI, NIH
    Bldg. 37, Room 6D24
    Bethesda, MD 20892

On 7-Aug-86, the total inventory of NSC-600000 was 800 mg, in one sample.

NSC-600000 is currently classified as 1D (Deferred: Does not meet DN-2 activity criteria), on the basis of its activity in P388 Leukemia.

The compound was tested in only one tumor system.

Figure 11.   An Executive Summary Produced from the DIS.

and it is becoming possible for computer systems such as the DIS to take much greater advantage of graphics.

Our ideas are still primitive - it is difficult to believe just how exploitable graphics are - but already it begins to appear, as we had suspected - that, faced with data presented in graphical form, humans can absorb it at astonishing rates, and with astonishing discrimination. If we can think how to express a megabyte of data graphically, the reader can absorb it with little thought. Examples abound (1) of graphics which, though they contain the equivalent of between 20,000 and a half million numbers on a single page, can be taken in at a glance by the reader. The same volume of data, expressed in alphanumeric form might spend months seeking a reader with adequate diligence, competence and stamina.

The lessons are simple: the technical difficulties associated with graphics are rapidly evaporating; the end-user, any end-user, functions extraordinarily well with graphics presentations; most end-users lack the attention span to deal with large amounts of alphanumeric output. The challenge that remains is to devise a clear, communicative graphical vehicle for your message. If you can do this, your computer and your reader can do the rest.

Literature Cited

1. Tufte, E. R. The Graphical Display of Quantitative Information; Graphics Press: Cheshire, CT, 1983; pp 16, 155.

RECEIVED April 10, 1987

Chapter 11

The Threefold Challenge of Integrating Text with Chemical Graphics

Robert M. Hanson[1]

Integrated Graphics, P.O. Box 401, Northfield, MN 55057

Three major challenges relating to the problem of integrating text with chemical graphics are detailed, and the approaches taken by the author in designing the FLATLAND system to meet these challenges are outlined. The FLATLAND system provides a model from which a new, inclusive definition of "integration" can be made, a definition based on flexibility and involving the entire process of document production.

The problem of integrating text and chemical graphics is readily apparent to any chemist: we get our graphics from many sources, for example, from instruments in the form of spectra and chromatographic traces, from databases in the form of chemical structures and reactions, from experiments in the form of synthetic pathways and kinetics data, and from our own imaginations in the form of mechanistic insights and hypotheses. We have the need to combine all of these various graphics with text to produce articles for publication, reports for internal communication, books, and theses. For some time word processors have been capable of dealing with most of the text-writing requirements of the chemist, but only recently have computer-based systems been developed which aid in the graphics area of the process.

Consider the task of putting together a document relating to chemistry, such as a thesis, a report, or an article for publication. What would it be like if we could take any table of data, any chart, any picture of a spectrum or molecule, or any chemical scheme or figure and, as we write, integrate that figure or table into our document? What if we could introduce new raw data, redraw a figure, or update the yield of a reaction in a scheme without complicated computerized cutting and pasting? What if we could do all this and not be limited to using a specific word

[1]Current address: St. Olaf College, Northfield, MN 55057

0097-6156/87/0341-0120$06.00/0

processor? What if we could integrate into our document raw data
from graphics, non-graphics or data base sources already in service
without having to write complicated graphics output interface
routines for each? Clearly if all this were possible, document
production would be considerably simplified in many ways.

The future for integrated graphics/document processing will
depend upon how well program developers respond to three basic
challenges. First is the challenge of developing a system which can
collect raw data from a variety of sources and transform those data
into graphic form. Second is the challenge to develop a system
flexible enough to operate with a variety of word processors,
graphic input devices, and printers, not only those available now,
but those of the future. Third is the challenge to develop a system
which makes possible the integration of the whole process of
document production, allowing rapid access to all of the stages of
production at any time. We will outline these three challenges in
turn and show how the FLATLAND system addresses each.

Challenge I: Multiple Graphics Sources

Chemical graphics come from a wide variety of sources. Getting this
diverse information into one document invariably requires a
substantial amount of "cutting and pasting". A major challenge
facing those of us interested in no-cut, no-paste document
production is how to get information, such as data from
spectrometers, structures from data bases, graphs from data
analyses, and schemes from synthetic work, all in a form suitable
for insertion into a document. What is needed is a general purpose
"black box" processor which can convert graphic data from sources
such as ORTEP (X-ray crystallographic analysis), MM2 (molecular
modeling), MACCS (Molecular Design Limited), and R/S 1 (BBN, Inc.;
for data analysis) into the form required by word processors. Three
approaches to this challenge will be outlined.

One approach to this problem, taken by Talaris, Inc., for use
with their laser printers, merges graphics and text files at the
final printing stage (QDRIVE). One of QDRIVE's principle strengths
is its capability to accept graphics data in several formats,
translating it on the spot for printing on the Talaris laser
printer. Thus programs which were originally written without device
independence in mind and producing output for Tektronix or Versatec
plotters require no changes for their data to be included in a
document.

Apple Computer's Macintosh system uses a radically different
approach. The Apple system employs a standardized device language
(Postscript) for all of its graphics, so any application
automatically creates graphics images in the proper format for
document inclusion. Apple has made it relatively easy for the
sophisticated user to write applications producing document-ready
graphics. The Apple system suffers only in that it is very
expensive on a large scale and cannot utilize graphics data from
more traditional sources, such as MM2 and R/S 1.

The FLATLAND system provides a model, on the other hand, for
how one graphics system can, in principle, utilize the raw data from
a variety of information sources, including operator input, public

and private structure and scheme libraries, molecular mechanics programs, X-ray crystallographic data, and in-house data bases.

Challenge II: Device and Word Processor Flexibility

New and more powerful word processors are becoming available each year. Getting graphic data in a form flexible enough to be used by more than one word processor is an ever-present problem. The problem is exacerbated by the fact that several laser printer systems exist, or are under development, each of which requires radically different protocols and graphic languages. To complicate matters, laser printer technology is not likely to stabilize for at least a few years. In order to be compatible with future word processing systems, graphics-producing systems such as FLATLAND must be designed to work with several word processors and generate output suitable for more than one output device.

Challenge III: Process Integration

One of the most frustrating aspects of word processing is that system designers have generally not considered the possibility that their product might be part of a larger system. Thus, word processors are not designed to allow easy communication with other concurrently running programs. The actual process of combining text and graphics information, however, generally involves going back and forth between writing of text and preparation and modification of graphics. Although the Apple system displays text and graphics together, fairly complicated deletion/reinsertion steps must be carried out when changes are required in the graphic part of the document. Such complications can be lessened by using switching programs (which allow multiple systems to be quickly accessed).

Most word processors do not merge graphics and text until print time, and thus do not allow for the ready visualization of graphics with text until all is on paper. The advantage here is that alterations in the graphics can be made entirely independently of the text, so no cutting and pasting is required. A clear challenge for word processors of the future will be to allow ready visualization and alteration of both graphics and text, with even computerized cutting and pasting being unnecessary. The FLATLAND system uses a novel switching technique to allow immediate transfer between text processing and graphics production.

The FLATLAND System

FLATLAND takes its name from the book FLATLAND: A Romance of Many Dimensions written by Edwin A. Abbott in 1884. In his book, Abbott depicts his world through the eyes of 19th century Victorian satire, imagining a land of two dimensions. Likewise, the depiction of our three-dimensional world of chemistry in the two dimensions of the printed page is the aim of the FLATLAND system, developed by the author for use on a VAX computer operating under the VMS environment.

Essentially, FLATLAND is a drawing program, capable of producing high quality structures and reaction schemes for output on

various video terminals and hard copy devices (see Figure 1).
FLATLAND, though, is much more than a two-dimensional drawing
program in that the third dimension is still there (see Figure 2).
FLATLAND is unlike other drawing programs in that information is
stored in <u>molecular</u> form, that is, as atoms, groups of atoms, and
bonds, rather than as lines and circles and letters. The data format
is an enhanced version of that used in molecular modeling, making
FLATLAND uniquely suited for interaction with data bases and
libraries of structural information. Data bases such as those used
by the MACCS system might be directly tapped for molecular
structures needed during the drawing session.

FLATLAND also has its own structure and structure fragment
libraries (see Figure 3). Structures and reaction scheme "templates"
can be used over and over while drawing to add whole predrawn
molecular or scheme sections to the developing scheme. The
libraries, both public (system wide) and private (user owned), are
dynamic and can be customized easily and continually.

FLATLAND Organization

The organization of information in FLATLAND is unique (see Figure
4). Documents are seen as a combination of text and graphic scheme
"pointers". A pointer is simply a one-line command indicating to the
word processor the space allocation (in lines) to be allocated for
the given scheme along with the name of the picture file (in printer
language) to be merged with the text at print time. FLATLAND
produces that picture file from a file called the <u>scheme file,</u>
which, like a document, includes text along with a set of pointers
to other files (in this case to <u>structure files)</u>. Structure files
contain lists of point positions, labels, and connections (atoms,
groups, and bonds). Both scheme and structure files may also contain
searchable, nonprintable comments.

Though hierarchical, FLATLAND's design allows for substantial
flexibility. Just as schemes can be modified entirely independently
of the document text, structures can be modified entirely
independently of the scheme. In addition, a single structure can be
used in many schemes without redrawing or copying, since each scheme
merely contains a reference to that structure. Schemes are device
independent and can be converted to picture files for output to
terminals, plotters, or laser printers. Picture files can then be
displayed, plotted, printed, or merged with a document.

FLATLAND and Challenge I

As mentioned above, a major challenge for chemical word processors
is the requirement to deal with graphic information from a variety
of sources, particularly on-line services, data bases, and
structure/reaction libraries. FLATLAND finds itself perfectly
situated between these sources and the word processor. No full-
capability word processor is likely to be developed which can
directly tap multiple, specialized graphic-oriented data bases.
Likewise, the often used programs SHELLX (for X-ray crystallographic
analysis) and MM2 (for molecular modeling) were not designed with
graphic manipulation and word processors in mind. Both are excellent

CHP/3A sieves
(+)-DET
OH
$Ti(O^iPr)_4$
CH_2Cl_2
O
OTs

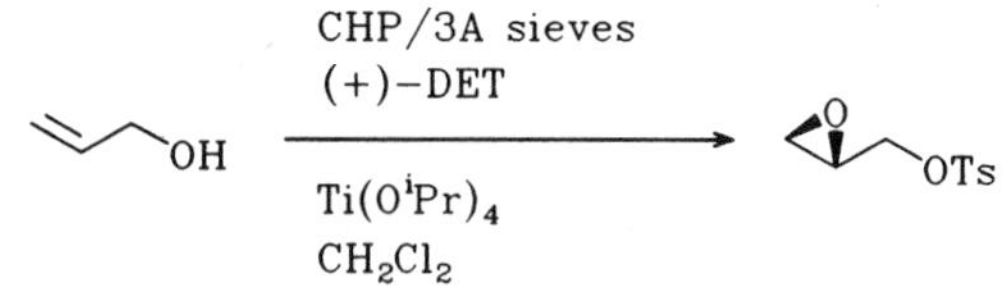

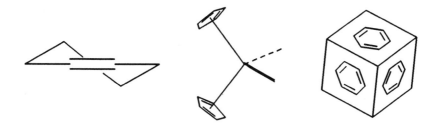

Figure 1. An example of a FLATLAND scheme.

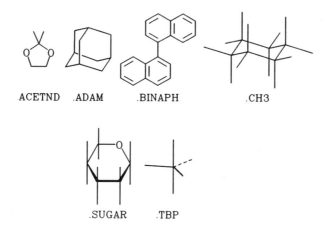

Figure 2. Three-Dimensional FLATLAND structures.

O
O
.ACETND
.ADAM
.BINAPH
.CH3
O
.SUGAR
.TBP

Figure 3. Examples from the FLATLAND library.

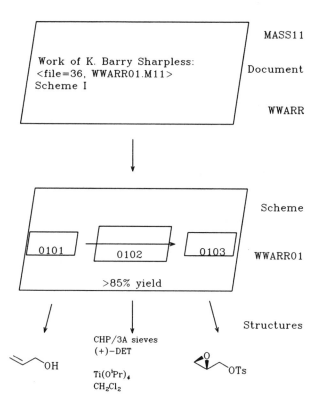

Figure 4.   FLATLAND organization   and the relationship between
FLATLAND schemes and MASS11 documents.

at calculating molecular geometries, but neither is especially good
at graphically presenting those geometries in ways that can be
easily integrated into documents.

FLATLAND's hierarchical design allows for structural
information from virtually any source to be utilized at scheme
production time. FLATLAND's design provides for automated run-time
access to any number of small, user-written reformatting/searching
programs. Thus, while preparing a scheme, one might retrieve a
structure from a previous (or concurrent) MM2 modeling session,
reading it in as a FLATLAND structure. That structure could then be
independently modified, three-dimensionally rotated, and rescaled
anytime prior to or after inclusion in the scheme. In fact, FLATLAND
could be used as a graphics "front end" to manipulate calculational
data three-dimensionally, thus supplying a graphics module to any
calculation or data base access system. Such use requires no changes
to the FLATLAND system whatsoever. All that is required is the
writing of a few small, independent data format translation
programs. Such programs could be quickly written by an in-house
system programmer.

FLATLAND and Challenge II

The Challenge of the ever-changing word processor scene is also
addressed by FLATLAND. Since schemes are not the actual pictures,
they can be device- (and thus word-processor-) independent. Schemes
are seen as the intermediaries, the organizers of information,
rather than the information itself. Thus, all information is
organized in FLATLAND without any dependence upon destination. The
same scheme can be converted to a displayable image, sent to a pen
plotter, or translated into a laser printer file for either direct
printing or inclusion in a document. The key here is that as new
devices become available, new device "drivers" can be added; as new
word processors are employed, the schemes can be "repackaged" to
remain compatible.

FLATLAND and Challenge III

FLATLAND is the first chemical graphics system ever designed
specifically to work concurrently with other systems. Information
can be transferred back and forth between systems, and FLATLAND can
be instructed to wait for other processes to call for graphics.

For example, let us say FLATLAND is being used to create
graphics for inclusion in a SCRIBE (Unilogic, Inc.) text. SCRIBE is
a document production system that relies on a system editor such as
EDT for text entry. SCRIBE is not a "what-you-see-is-what-you-get"
system by any means. Rather it is a very smart processor, which can
automatically move blocks of text to fit page boundaries, handle
multiple numbering systems (tables, figures, compounds, and
references, for example), and insert picture files from external
sources (FLATLAND, for example). All SCRIBE needs in order to insert
a picture is a simple command, for example @pic(size=36 lines,
file=WWARR01.LNO), stating the size of a picture in lines and the
VAX filename associated with that picture.

The novel aspect of FLATLAND is that both processes, text writing with SCRIBE and scheme production with FLATLAND, can be carried out concurrently on the VAX. When a scheme is ready, one simply jumps to the text and presses a key. The proper SCRIBE command for that scheme is placed directly into the text where it is needed, including the calculated space allotment. No measuring is necessary. Note that the scheme is not actually "placed" into the document; only a reference, a "pointer", is introduced. Using MASS11 (Microsystems Engineering) the procedure is similar, but in that case the "embedded" command referring to the picture, <file=36, WWARR01.M11>, must be entered manually. FLATLAND notifies the user of exactly the format to use, again including the calculated space allotment. In either case, the effect is an integration of the processes of producing the schemes and writing the text of a document.

Summary: Integration and FLATLAND

In summary, integration in the context of FLATLAND takes on a much more encompassing definition than has previously been given the term. Integration is not seen as merely the placing of graphics data with text data on the way to a laser printer. Nor is integration seen as providing a "what-you-see-is-almost-what-you-get" view of a finished document. Rather, integration is seen as the bringing into a single operation the entire process of putting a document together. Integration is seen as the capability to draw upon graphic information from a wide range of sources at document production time, and to make possible the concurrent production and modification of both text and graphics.

It is hoped that developers of word processors of the future will recognize that no system can "do it all". The key lies in establishing a division of labor: in letting modeling programs such as MM2 and data base access systems such as MACCS provide the structures, in letting graphics programs like FLATLAND provide the link, and in letting processors such as MASS11 produce the text. Only then will we have complete integration; only then will we have a flexible system allowing immediate access to all stages of document production.

RECEIVED March 25, 1986

Chapter 12

Computer Graphics at the American Chemical Society

Jack M. Sanderson and David L. Dayton

Chemical Abstracts Service, American Chemical Society, P.O. Box 3012, Columbus, OH 43210

The American Chemical Society (ACS) publishes 19 Primary Journals, Chemical Abstracts, and many other publications. These publications are an information source for ACS electronic data bases available from the Scientific and Technical Network (STN International) and from other on-line vendors. The ACS currently processes computer graphics of chemical structures for publication and for display through STN International, both on- and off-line. The ACS is also involved with development of graphics software to automate the display of images for the U.S. Patent and Trademark Office and to load a file of German patents containing line drawings onto STN International. The computer systems that provide chemical structures for the ACS publications and for STN International with the Messenger Software are described. Techniques are presented which will be implemented to handle a wider range of computer graphics, not limited to chemical structures.

The American Chemical Society (ACS) publishes 19 primary journals, Chemical Abstracts (CA), computer-readable and on-line data bases, and many other publications in chemistry. Chemical Abstracts Service (CAS), a division of the ACS located in Columbus, Ohio, does most of the computer processing and software development for these publications and data bases. ACS publications and data bases contain chemical structures, tables, equations, line drawings, and other images. CAS continues to automate the processing of graphics for both printed publication and on-line data base publication.

CAS cooperates with Fachinformationszentrum Energie, Physik, Mathematik (FIZ-Karlsruhe) in Karlsruhe, Federal Republic of Germany, and the Japan Information Center of Science and Technology (JICST) in Tokyo, Japan, to form STN International, the scientific and technical information network. The CAS-developed Messenger software is used to support STN International and is running on

fonts of line segments and special characters that are used for
drawing chemical structures. Text and chemical structures are
formatted together (both are represented only by characters) on the
mainframe and sent to the Xerox 9700 which prints at its maximum
rate of two pages per second.

The ACS primary journals are produced by a batch mainframe
system developed by CAS over 12 years ago. Some limited graphics
capabilities have since been added for tables and equations.
Tables and equations are input as text (not graphic) entities
associated with a manuscript. Tables are keyboarded in a form with
each data element tagged, somewhat like the input conventions for
"tbl" processing available with UNIX. A batch format program
transforms the table data elements into an image suitable for
publication. Table modification is done in batch mode by changing
the tagged representation, rerunning the format program, and
producing typesetter output for review and possible correction.
Chemical and mathematical equations are processed by an on-line
system adapted from the OLSIS system. Equations are keyboarded as
a text string of special characters with baseline or font changes.
Another mainframe program transforms the string into a formatted
equation and stores it in the CAS standard of GDS. After creation,
the equation image is displayed on-line at a Sanders 7 graphics
terminal along with the original text string. For on-line editing,
the text string is modified, the format program is rerun to produce
a new GDS, and the new equation image is displayed on the graphics
screen. The only graphics portion of the ACS primary journal
production is this on-line display of the equation image stored in
GDS.

Tables and equations are handled as graphics elements defined
by a "window" size for placement on final pages. Other line art
and images have not yet been automated and are stripped-in by hand
on final pages. However, the measurements for image windows are
entered into the primary journal data base via the on-line equation
system so that the proper amount of white space is reserved on
final pages. CAS has developed a sophisticated batch pagination
program which composes text and graphics elements for entire
primary journal issues, including tables of contents and indexes.

A problem CAS intends to address is that different software is
now used to handle chemical structures for different purposes.
Ideally, one set of software would handle chemical structure input
and storage serving the multiple purposes of substance
registration, publication in ACS services and in STN International,
and structure query input to STN International. The emerging
graphics standards and advances in graphics tools will help CAS
solve the problem.

Graphics Standards and the CAS Storage Mode: GDS

A graphics device either inputs or outputs graphical commands, or
does both. These devices are controlled by a software component
called a "device driver" that translates commands from a graphics
application representation to output primitives for a specific
graphics device, and vice versa if the device can be used for
input. Because of different device technology, device drivers are

device-dependent. Almost every hardware vendor has a different
means to access graphics devices. Software developed for one
machine is usually not very useful on another. If a new graphics
device is added or changed, large parts of the application software
have to be reprogrammed. Until graphics standards are used
uniformly, application programs will still have to be developed and
maintained for specific device drivers. Standards are aimed at
enabling graphics applications to be developed which are portable,
running with a variety of graphics hardware.

Historically, there have been many forms for representing
graphics, most derived from a specific vendor's implementations.
The situation is changing. Emerging graphics standards may promote
development of device independent graphics application software and
may eventually lead to compatibility of graphics devices. In this
section, the general features of GDS (the CAS storage form) are
described along with the other adopted and de facto standards.
Comparisons and contrasts will be drawn from the characteristics of
these standards and representations, and the advantages and
disadvantages of standards explained.

Standards are being developed for the two main graphics
interfaces: the graphics applications interface and the device
driver interface. The programmer interface is the set of graphics
application functions and syntax that allow a programmer to access
standard graphics utilities. The application interface standard
provides a common syntax between the graphics function calls and
the device driver, see Figure 1. Protocol A represents the
application interface standard. The device driver interface
standard, Protocol B in Figure 1, specifies I/O commands (e.g. line
drawing primitives) that are the same for all graphics devices.

Examples of graphics applications interface standards are the
Graphics Kernel System (GKS) (6) and the Programmer's Hierarchical
Interactive Graphics System (PHIGS) (7). Specific interface
definitions are part of both GKS and PHIGS. For the device driver
interface, examples are the North American Presentation Level
Protocol Syntax (NAPLPS) (8) and the Computer Graphics Interface
(CGI) (9).

The CAS-designed Graphical Data Structure (GDS) is used
exclusively to represent all graphical images in storage. It is
the CAS internal standard. There are two forms, one optimized for
efficient manipulation by software and the other for minimizing
storage. The two forms are logically equivalent and CAS has
developed conversions between them. CAS has also developed
mainframe foundation routines to build, manipulate, and update GDS,
and to translate GDS into forms suitable for output on the APS-5
typesetter, the Xerox 9700 laser printer, and selected graphics
terminals (e.g. the Tektronix 401X).

GDS is a hierarchical tree structure consisting of the root or
head node, intermediate nodes, and terminal nodes. Branches
connect nodes; there may be several branches leaving a node, but
only one branch entering a node. The machine representation of GDS
provides pointers between these elements and maintains the
hierarchy in a form that can be efficiently manipulated. The root
of the tree by definition has a (X,Y) coordinate of (0,0) at the
origin of the display surface on the device or graphics terminal

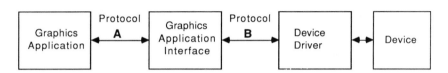

Figure 1.   Interfaces Between the Graphics Applications and Devices

screen. Branches specify displacements in position relative to the
previous node. (X,Y) values are specified in intermediate nodes as
deltas or changes of position. Terminal nodes specify picture
elements which will be drawn at the position calculated by summing
the displacements while traversing the tree from the root or origin
of the display screen. Elements in terminal nodes may be vector
line segments, arcs, enclosed areas, characters, or other elements.
Currently, only two-dimensional (2D) images can be stored in GDS.
Parameters are specified at appropriate points in the tree
structure for different devices and their characteristics.

The main points are that GDS is hierarchical, uses relative
coordinates, and has a variety of picture elements. It is not
specific to chemical structures. Conveniently, GDS graphics
elements are primitives similar to the GKS graphics standard. In
the future, CAS could develop an automated software conversion
between them.

The recently approved international standard for 2D graphics
is GKS. GKS specifies standardized programming methods and
interfaces for application programmers. Thus, GKS achieves
application code portability and provides a standard interface to
device drivers. For long-term storage and transmission of
graphical data, a GKS storage form (the GKS Metafile) is provided
by the standard. Also, a workstation description standard
specifies a table that describes the capabilities of a device. The
table is analogous to UNIX "termcaps" (10) for defining text
terminal characteristics.

Six output primitive types are available with GKS: polyline
(generates a connected series of points), fill (shades in a
polyline), polymarker (generates a symbol at a given point), shape
(generates arcs, rectangles, and such), array (generates cells with
different colors), and text (generates a character string at a
given point). Note that these functions are similar to those of
GDS, the CAS standard. Several companies have developed GKS
subroutine libraries for the FORTRAN, C, Pascal, and BASIC
programming languages. An extension to GKS supporting
three-dimensions (3D) has been proposed.

PHIGS is the other leading application interface standard
being proposed. Many of the PHIGS basic functions, such as output
primitives and logical input devices, are identical or similar to
corresponding GKS functions. The major differences provide PHIGS
with file hierarchy, image "substructures," and the ability to
modify dynamically the relationships of substructures in the
hierarchy. PHIGS is designed for sophisticated manipulation of
graphics data and 3D representations, as it addresses the graphics
needs of scientific and engineering applications. It is possible
that GKS and PHIGS will merge, but it is also possible that PHIGS
will become a separate standard from GKS, useful for heavyweight
graphics applications.

The most general device driver interface is the Computer
Graphics Interface (CGI, called "VDI" in Europe). CGI will
standardize device drivers, making all graphics devices appear to
be identical by defining an interface to a virtual (ideal) device.
This standard interface protocol is converted to the actual form
for a real device in the host (with a device driver) or in the
graphics device itself.

Standardized software tools are available to support building device drivers and IBM is supporting CGI. CGI could allow the implementation of a generalized device driver in hardware chips, thus offloading host processing to the graphics terminal or intelligent workstation. If implemented, this would provide true hardware independence.

NAPLPS was designed specifically to meet the needs of videotext on slow speed telecommunication lines; currently, it is the videotext standard in the United States and Canada. As a communication standard for the exchange of graphical and textual information, it is a useful device interface because it is compact, portable, and resolution independent. A decoder is needed on the terminal to receive and display NAPLPS. This can be a software emulator or it can be built into the hardware.

NAPLPS is a superset of ASCII with extensions that allow for display of enhanced alphanumeric text and images. Words and images can appear on the same screen. Display is accomplished through the use of operation codes that tell the terminal which predefined or variable (dynamically redefinable) character sets to use, which graphics shape from a repertory of simple geometric primitives (point, line, arc, rectangle, or polygon) to draw, what kind of environment to use to display the graphics and text on the screen (color, blinking, stroke width, textures, etc.), and where to draw them.

CAS does not plan to use NAPLPS, because it is not as versatile as other device driver interfaces. NAPLPS is not conducive to manipulation of complex images, such as chemical structures. However, it is feasible to use it as a display format, much as STN International uses PLOT-10 today. Wider acceptance of videotext in other on-line applications (e.g. catalog display and order entry) is needed so that more people will have NAPLPS terminals and decoders. For graphics applications which serve a large customer base, it is important to provide support for the terminals that customers have. STN International customers are unlikely to have NAPLPS terminals, so it is not currently considered a graphics representation which needs to be supported.

Page Description Languages (PDLs) (11) are a recent advancement in representing text and graphics for 2D output. PDLs are designed to specify the exact layout of a page image formatted for output, including text in various sizes and fonts, graphics, and digitized images. PDLs are really programming statements to be interpreted and dynamically executed on the output device. PDL output devices are equipped with a computer chip that processes the PDL statements and builds a digitized image in the bit-map memory on the output device. These chips are called RIPs or Raster Image Processors. Once built, the page image is transferred directly from the bit map memory to the display or print engine for output. PDL statements represent graphics and text in a compact and resolution independent form. Like GKS, PDLs have primitives capable of producing any 2D graphic. However, unlike GKS, once a page representation is cast in a PDL, it is useful only for display purposes and cannot be manipulated easily. PDLs are exciting because resolution independent output can now be sent to a wide variety of output devices capable of producing graphics and text.

The next logical step is for PDL chips to appear on graphics terminals with built-in text fonts, thus providing another alternative for screen display.

The leading examples of PDLs are PostScript (12) from Adobe, and Interpress (13) from Xerox. Another late arrival is DDL from Imagen. PostScript, for instance, can be sent to over 20 different models of laser printers or photocomposers from different vendors, producing output at resolutions from 300 to 2000 dots per inch.

A comparison of standards is difficult and uses must be tailored to specific applications. As CAS discovered early, both efficient manipulation and compact storage must be provided. This is especially true for large data bases, such as those on STN International. I/O to a variety of devices must be supported. Since STN International customers have graphics terminals and communicate over slow speed lines, CAS would like to support a variety of graphics devices with a representation that is compact enough for transmission over slow speed telecommunication lines in a reasonable amount of time. CAS has yet to complete analysis of this problem, but it is probably not possible to meet all these goals without making some tradeoffs.

Because of the advantageous mathematical properties of vectors, CAS has determined that line drawings should be stored as vectors and other primitives, rather than as digitized images. Graphics represented by vectors can be more easily manipulated, can be more efficiently stored, and can be more quickly transmitted than compressed digitized images.

Critics argue that GKS (and other standards) do not provide the full set of graphics functions available on all devices and that inadequate performance will result. However, GKS is suitable for all but the most sophisticated graphics modeling applications. It provides an impressive set of features that will meet present and near-future CAS needs. However, it cannot always be implemented with adequate performance, and the addition of new features, such as 3D, may cause applications using it to be even slower. Even though CGI is not an official standard yet, software vendors are introducing early implementations of it and future announcements deserve to be closely watched. Graphics hardware vendors may not move as quickly as software vendors, nor wholeheartedly support a standard that makes their products more compatible.

Despite some drawbacks in using graphics standards, CAS expects to gain benefits in software portability, programmer productivity, and device independence. Graphics applications developed for one system can readily be moved to another. Software developers can concentrate on applications development instead of learning how to use new graphics implementations. With device independence, a wider variety of graphics devices can be used without application software changes. This has the potential to open up possibilities for supporting a wider variety of graphics terminals on STN International in the future.

Potential Future Graphics Applications at the ACS

GDS (the CAS internal standard) is concerned with the creation, manipulation, and storage of graphics. GKS and PHIGS are concerned with displaying graphics and returning changes made to displayed images. CGI is concerned with standardizing the graphics interfaces to devices. PostScript provides a versatile method for displaying formatted graphics and text. CAS feels that these standards and technology advancements can work effectively together. Figure 2 shows a short-term scenario in which CAS develops future mainframe graphics applications using the expertise and tools already in place for GDS, while taking partial advantage of GKS and CGI. The cornerstones for maintaining compatibility with emerging standards are for CAS to develop applications using GKS toolkits and to develop a PostScript driver for GDS. It would also be possible to develop a PostScript driver for GKS, but CAS will wait to see if one will come from a commercial software source. Implementations of window manager software utilizing PostScript have already been announced and CAS is considering them for use in workstations. In the long-term, GDS could be replaced with a storage mode standard, such as the GKS Metafile. PostScript will be useful for display purposes, both on laser printers and eventually on graphics terminals. PostScript or GKS Metafile may be future alternatives for softcopy distribution.

FIZ-Karlsruhe, with assistance from CAS, is developing a prototype system for displaying line drawings from STN International files. Figure 3 is a line drawing that demonstrates the capabilities. The scanned image is captured at about 400 dots per inch. The digitized image is converted to vectors by a software algorithm, achieving a large compression factor of the data while maintaining the original resolution. The vectors are in a polyline form, i.e. a series of short line segments enclosing areas to be shaded. The vectors are converted to ASCII characters and then transmitted to a workstation for display. Figure 4 shows the data path envisioned for sending line drawings via STN International to an intelligent workstation for display.

The workstation used is an IBM/PC AT and transmission of the image takes slightly over 1.5 minutes via a slow speed 1200 bits per second line. The image vectors are displayed on the workstation with a graphics card driven by software to convert the vectors back to pixels. These vectors can be converted to PostScript at the workstation and sent to an attached laser printer for hardcopy output at 300 dots per inch. The picture used as Figure 3 was made from the actual laser printer hardcopy output. It is also feasible for the vectors to be converted to and manipulated in the GKS or PLOT-10 vector format, but this has not yet been done.

With further technical improvements, such as faster data transmission lines, the prototype may make it possible to display images of line drawings, not just chemical structures, from STN International files. Images would include spectral data, chemical apparatus, graphs, charts, mechanical drawings, and other line drawings.

The ACS primary journal system may be upgraded to handle automatically structures, line drawings, and scanned images. This would provide for both publication of the images and their eventual inclusion in the full-text CJACS file recently loaded on STN International.

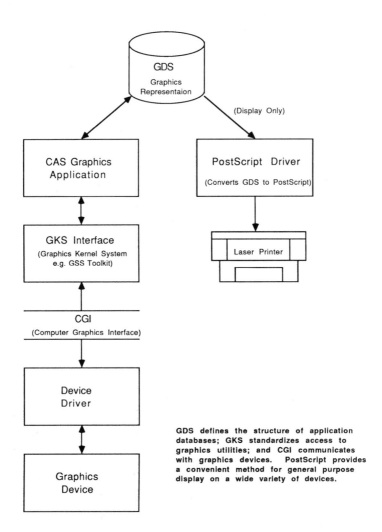

GDS defines the structure of application databases; GKS standardizes access to graphics utilities; and CGI communicates with graphics devices. PostScript provides a convenient method for general purpose display on a wide variety of devices.

Figure 2.   Possible Architecture for Future CAS Applications

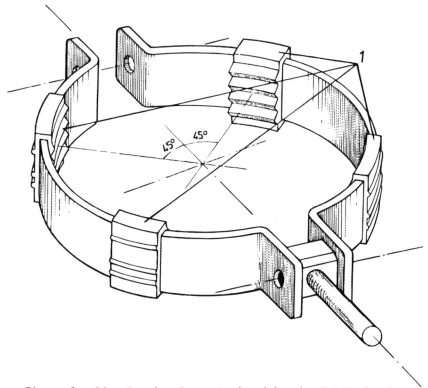

Figure 3. Line Drawing Image Produced by the FIZ-Karlsruhe
Prototype System

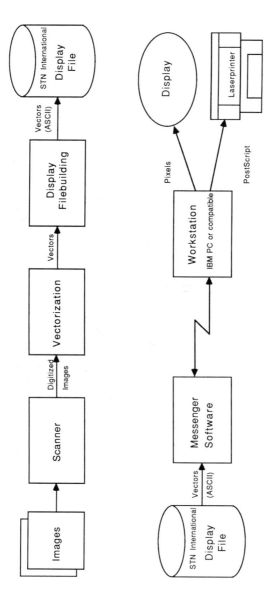

Figure 4.   Data Flow for the FIZ-Karlsruhe Image Display
Prototype

CAS will develop workstation software to build the Markush structures for a new Patent Service. The graphics software for building chemical structures could, ideally, also be used to build structure queries for STN International, connection tables for registration, and an image that is the appropriate quality for ACS publications.

CAS is developing software for the Automated Patent System (APS) at the U.S. Patent and Trademark Office (USPTO). Patent text, compressed digitized images, and associated reference files will be stored on a distributed network of file servers accessing optical disk jukeboxes. When loaded, the full set of files is expected to take 30 terabytes or 30 trillion characters of storage. A high speed local area network will link host computers, file servers, and specialized graphics workstations. Images will be captured at 300 dots per inch and displayed at 150 dots per inch (full-sized) or at 300 dots per inch (zoomed). Messenger software (on mainframes) and backend search engines (minicomputers) will do full text searching on post-1970 patents, and on USPTO classification index or patent numbers prior to 1970. CAS may gain experience from automating the USPTO that helps in developing new ACS graphics and image applications.

Summary

CAS intends to keep abreast of emerging graphics standards in order to integrate standards into our applications, or at least remain compatible with them. The ACS will use standards to promote the electronic exchange of information and potentially offer better service to STN International customers, such as perhaps displaying line drawings and supporting a wider variety of graphics terminals.

Literature Cited

1. Dittmar, P. G.; Stobaugh, R. E.; Watson, C. E.   J. Chem. Inf. Comput. Sci. 1976, 16, 111-121.
2. Farmer, N. A. et al. J. Chem. Inf. Comput. Sci. 1983, 23, 93-102.
3. Dittmar, P. G.; Mockus, J.; Couvreur, K. M.   J. Chem. Inf. Comput. Sci. 1977, 17, 212-219.
4. Blake, J. E.; Farmer, N. A.; Haines, R. C. J. Chem. Inf. Comput. Sci. 1977, 17, 223-228.
5. Farmer, N. A.; Schehr J. C. Proc. Assoc. Comput. Mach. 1974, 2, 563-570.
6. GKS: Standard X3.144, may be ordered for $33 from: CBEMA (The Computer Business Equipment Manufacturers Association), The ANSI Sales Dept., XE Secretariat, 311 1st Street, Suite 1200, Washington, DC 20001.
7. PHIGS: A Working Document, may be ordered for $35.00 from CBEMA.

8. <u>NAPLPS: Standard X3.110</u>, may be ordered for $35.00 from CBEMA.
9. <u>CGI: Standard X3.122</u>, may be ordered for $35.00 from CBEMA.
10. <u>TERMCAP(5), UNIX Programmer's Manual</u> May 10, 1980, 4th Berkeley ed., 1-8.
11. <u>The Seybold Report on Publishing Systems</u> 1986, <u>15</u>, 11-16.
12. <u>PostScript Reference Manual</u>; Adobe Systems Corporation: Reading, MA, 1985.
13. "Interpress Standards Catalog" may be ordered from Xerox Systems Institute, Sunnyvale, CA 94086.

Bibliography

Brannigan, M. <u>Computer Language</u> May, 1986, <u>3</u>, 26-30.
Clarkson, I.; Skrinde, R. <u>UNIX World</u> 1985, <u>2</u>, 47-48, 50+.
Clarkson, T. <u>Computerworld</u> April 22, 1985, <u>19</u>, 13ID-14, 16.
Heck, M. <u>Computer Graphics World</u> January, 1986, <u>9</u>, 49-50, 52+.
Holland, G. L. <u>EDN</u> January 10, 1985, <u>30</u>, 179-186, 188+.
Hurly, P. <u>Videodisc and Optical Disk</u> September-October, 1985, <u>5</u>, 372-387.
McCune, D. <u>PC World</u> April, 1985, <u>3</u>, 178-187.
Myers, E. D. <u>Datamation</u> May 1, 1986, <u>32</u>, 48-52.
<u>The Seybold Report on Publishing Systems</u> April 16, 1986, <u>15</u>, 11-16.
Paller, A. <u>Computerworld</u> April 17, 1985, <u>19</u>, 12-16.
Plunkett, B. <u>Computers in Mechanical Engineering</u> March, 1985, <u>3</u>, 37-40.
Rhein, B. <u>MIS Week</u> November 6, 1985, <u>6</u>, 41.
Rhein, B. <u>MIS Week</u> August 21, 1985, <u>6</u>, 40, 44.
Schnatmeier, V. <u>UNIX/WORLD</u> September, 1986, <u>3</u>, 26-41.
Shipley, C. <u>PC Week</u> May 14, 1985, <u>2</u>, 113, 131.
Sullivan, K. B. <u>PC Week</u> May 14, 1985, <u>2</u>, 116, 118.
Williams, T. <u>Computer Design</u> December, 1985, <u>24</u>, 38-39.

RECEIVED March 25, 1986

Chapter 13

Integration of Technical Drawings in a Data Bank System

Walter Niedermeyr[1]

Fachinformationszentrum Energie Physik Mathematik GmbH,
D-7514 Eggenstein-Leopoldshafen 2, Federal Republic of Germany

Technical drawings are indispensable data in presenting,
explaining, and transferring technical and scientific
information. Patent applications and utility models are
a field in which the illustrative and explanatory
function of drawings is particularly important. This
paper describes the joint storage of text and drawings in
one database, the conversion of digitized graphical data
into vector graphics output format and the combined
transmission of text and graphics via telecommunication
networks to various types of terminals. The advantages
of vector graphics over raster graphics for storing,
transmitting and displaying technical drawings are
discussed.

The project "Deutsches Patent- und Fachinformations-system" is an
attempt to make all the important information elements contained in
a patent document accessible via an on-line databank service. This
means all information in a document has to be transformed in such a
way that it can be transmitted through public networks using ASCII
7-bit code. Textual databases present no difficulties in
transferring data, also with an extended character set, because of
a great variety of solutions offered to get the information through
the networks to the end-user. Even chemical structures up to a
certain complexity can be represented by special use of given ASCII-
characters and symbols. This was one of the reasons that on-line
information systems based on chemical sciences could provide the
users with almost all the information contained in a document.
Other natural sciences and technical sciences are prevented from
applying such techniques, because in almost all cases there is no
possibibility of adapting technical drawings, curves etc. by means
of a combination of ASCII symbols. We investigated the
possibilities of adapting drawings with an extended pool of symbols,
available in a videotex environment. For reasons of principle which

[1]Current address: Gesellschaft für Mathematik und Datenverarbeitung,
Postfach 700363, D-6000 Frankfurt 71, Federal Republic of Germany

are not treated in depth here, this technique must fail. The most
important obstacles are resolution problems, which can only be
avoided by using a pixel representation of the drawing in the range
of 300 dots per inch and more. Scanning of the drawings and all the
other information which cannot be represented by standard ASCII
characters is therefore mandatory. Like all the other data stored
and processed in and recalled from a computer, pixels must be coded
in bit form. For example drawings are broken up into pixels of a
raster image, and each element of a drawing is assigned a bit
defining it as belonging to the drawing (black) or the background
(white).

Integation of binary raster images of line drawings in a databank is state-of-the-art and problems related to a pixel
representation of images (facsimile encoding) are well-known.
Particularly, compression of the voluminous data files is needed.

Facsimile Reproduction Techniques

Coding techniques for compression of binary images from the bitmap
representation have been developed for digital facsimile
reproduction (1,2). There are one-dimensional facsimile codes
grouping image elements of several lines.

Most one-dimensional image coding techniques group the image
elements of one line into runs of the same colour and indicate the
length of these runs in scanning order. This procedure is common to
all so-called run-length codes, while the coding technique itself
may differ. Although one run may comprise all image elements within
one line, short run-lengths are dominant in practice. This makes
fixed code length along the whole range of values impractical.
Commonly, run-lengths are classified into groups, with short runs
represented by a single code word and long runs by two code words.
Simple run-length codes use code words of fixed length while more
complex coding schemes use code words of variable lengths. Short
code words are assigned to frequent run-lengths and long code words
to less frequent run-lengths. A method of selecting code words was
first presented by Huffman, and coding schemes of this type are
referred to as Huffman codes. In general these coding schemes yield
more compressed data but require more complex coding and decoding
algorithms.

Two-dimensional facsimile coding schemes also make use of the
statistical dependence between lines of runs. In some coding modes,
runs whose initial or end points have shifted only slightly from the
lines before are processed as a block. The single coding modes have
different frequencies. More frequent coding modes are assigned
shorter code words than less frequent coding modes.

Finally, e.g., in the Japanese READ coding scheme,
superimposing coding modes may be used. Of the possible
alternatives the system will then choose the coding mode generating
shorter code words. This way, the line coherence of the original
image will be utilized and the coding scheme will be adapted in some
degree to the orginal document image.

With two-dimensional coding schemes, data transmission errors
will propogate through all follow-up lines of an image section coded
as a block. Short record lengths (blocks of two to four lines) help

to prevent this but will usually destroy the achieved data reduction
efficiency. Still, the data compression is two to three times
higher with these techniques than with one-dimensional Huffman
codes. Two-dimensional facsimile coding schemes are most efficient
in substituting data compression time for data transmission time.
The code words of two-dimensional facsimile codes are not directly
related to the geometry of the original image, so that geometric
operations (i.e., scaling operations) cannot be carried out.

Reproduction of Binary Raster Images

Complex coding and decoding algorithms are CPU-intensive operations.
Processing of drawings involves input of drawings of different scale
and output on displays and printers of different resolutions and
formats.
        To get a user friendly service, the principle "one drawing one
screen" must be achieved, i.e., output raster adaption is to be
provided. If possible, raster adaption should take place in the
receiving station. Raster adaption should be part of the image
decoding procedure as far as possible. Most of the available
devices for raster display of binary images (e.g. raster screens,
matrix or laser printers) today use bit map memories. Modern
personal computers make bit map memories accessible to the user via
suitable interfaces, using the bitmap memories both as refresh
memories and as main memories for graphical image manipulation (3).
        In recent times there has been a tendency to support higher
resolution in a PC, moving from 600 x 400 pixels to 1000 x 1000
pixels per screen at a tolerable price. However, due to the
enormous space requirements, bit maps are not suited for large scale
storage and teletransmission of binary images. For example, one DIN
A4 page with a raster of 16 lines/mm covers almost two Mbytes in bit
map representation.

Vector Images

As an alternative to the raster compression technique, there are
techniques reducing binary images to their geometric information.
The information contained in a picture is represented by broken
lines following either the center-lines or the outlines of the
drawings. These techniques are commonly referred to as
vectorization techniques as they mainly consist of position vectors
of the corner points of polygon lines.
        Vectorization techniques, especially when combined with
suitable techniques for broken line approximation in digital images,
yield a data compression similar to or higher than the most
efficient raster compression techniques. In contrast to compressed
raster images, vectorization techniques compress the image geometry
itself without breaking it up. This, in turn, means that input-
output raster adaptation for the procedures will require much less
time. Facsimile coded images must be transferred into rougher or
finer bit maps after decompression. The quality of this procedure
varies with the scaling factors used.
        Image vectors, on the other hand, group image points of
geometrical relevance. Rescaling is possible in the vector format,

and transfer of image vectors into a bit map of arbitrary resolution
for output and display is a simple operation.

The Vectorization Model of Transmission and Display of Text and Graphics in the Patent Database

When we started to produce a patent database with both text and
drawings, we were faced with the task of finding a mode of
representing patent drawings (without half-tones) assuring fast and
accurate electronic production of images from rasterized graphic
material. Continuing from our preliminary study concerning
representation, transmission and display of technical drawings on an
appropriate receiver station we decided to start working on the
vectorization of data, based on already scanned and digitized
original copies.

In the following, we concentrated on the characteristic
features and efficiency data of commercial systems and system
components for vectorizing binary images. The systems were tested
during visits to various US and European institutions. Principally
one can differentiate between two main vectorization techniques:
center line and outline representation.

Skeleton images are used, e.g., for digitizing geographical
maps. In cases where line widths are not important, center line
vectors will be sufficient (Figure 1a). To generate center line
points, the components of the initial images are broken down to
simple lines by continuous removal of edge points. Thinned lines
can be represented by chain (4) codes or by approximation.

The chain coding procedure requires only the coordinates of the
initial point of a line and the raster increments of its other
points. With line approximations of polygon lines, the polygon line
geometry is disregarded only within pre-determined (5,6) limits
(Figure 1b). Normally, the corners of approximated polygon lines
will not be on neighbouring pixels. They can be represented by a
suitable generalized chain code. Limits of line segments (Figure
1c) can also be represented easily by a further generalization of
the chain code, e.g., in an additional component of the corner point
coordinates.

Center line images are usually generated by binary thinning
operations on a bit map of the initial image. This procedure, which
requires considerable processing time on a general purpose computer,
can be avoided by outline vector imaging. In this technique, the
outlines of image objects are directly represented by closed lines
going either through the edge points of the image objects (Figure
1d) or following the cracks of the image objects (Figure 1e).

Outlines, like center lines, can be represented by chain codes.
Direct approximation of outlines is possible without line width
data. It is sufficient to distinguish between outer and internal
contours. Apart from the structural properties of the vector images
themselves, also the algorithmic properties of the image generating
procedures must be considered. Efficient approximation methods act
as geometric filters. They may be designed in such a manner that
standard increments of the initial lines are assumed and accumulated
into approximating increments. The processing time is constant for
all points on a line. Vectorization methods proper combine outline

center line points of a binary image into lines or planes. Due to
the large size of the bit maps of binary images, the efficiency of
these techniques depends on how access to the single image points is
organized.

The processing time is a function of the internal data management. Methods requiring searching of working areas need more search
time per item with increasing working file volume than methods
getting internal data structures in constant time. Most techniques
abandon this scanning sequence in certain image situations for line
vectorizing operations. This requires image map storage, either
whole or in parts. The increasing number of direct accesses makes
these techniques less efficient than strictly sequential coding
schemes. To avoid backtracing of lines, the image structures
already passed are retained in internal data structures until output
in the form of closed lines is possible. Other methods have a
single access for each image point in fixed scanning sequence.
Image map memories are not required, and vectorization takes place
immediately after image scanning.

The systems analyzed so far have given us some idea of the
requirement to be made on an ideal vectorization system for the
present project task. An optimum method of raster image
vectorization for storage in a databank and output on displays and
printers of different resolution should have the following features.
Outline images are to be preferred to center line images. They
avoid artefacts resulting from quantization of line widths and give
a faithful representation of solid area features. Mixed outline/
center line images still have the shortcomings of center line images
and in addition, are difficult to adjust due to the necessity of
tracking system parameters. Closed polygon lines have smaller
circumferences than trapezoids derived from them. Instead of one
closed polygon line, each trapezoid represents two separate outline
components which are more difficult to follow by subsequent line
approximation than closed polygon lines. Therefore we decided to
convert the original image into closed polygon lines following the
contours of the original image. All filled areas of a drawing are
represented only by their outlines.

In the IMAGIN company, we found a partner with excellent
knowledge in this field who was able to offer us the tailored
solution we use nowadays. The software product SCORE (Scan
Conversion for Outline REpresentation) was developed by the IMAGIN
company (7). The vectorization method meets all the requirements in
an optimum manner.

The vectorization technique is well founded mathematically.
Reduction of the number of corners of the polygon lines by
subsequent line approximation is controlled. The permissible
tolerance of this processing step is the only parameter of the whole
vectorization method.

The faithfulness of reproduction is directly related to the
tolerance of the line approximation method. The outline smoothing
is a vector prolongation procedure where the area deviation of the
new vector versus the old outline vector sequence is balanced (see
Figure 2) according to a given tolerance.

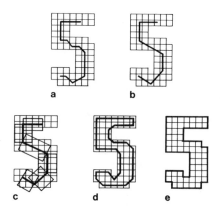

Figure 1.   Centerline and Outline Representation.

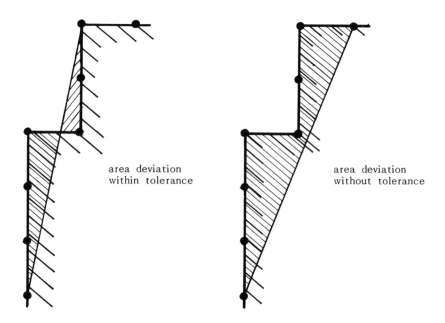

area deviation
within tolerance

area deviation
without tolerance

OUTLINE SMOOTHING = VECTOR GENERATION

Run length input is first converted to chaincode
and then smoothed to outline vector sequences

Vector components can be bound

Figure 2.   Outline Smoothing.

Testing the System

The maturity of SCORE was tested using 83 technical drawings from
about 4,000 patents available from the DPA (German Patent Office),
Berlin. The drawings were selected with the goal of getting the
whole spectrum of patent drawings for the final test; they were
classified according to complexity in order to obtain mean values
for storage requirements, vectorization times, and other parameters
of interest. Several drawings bordering half-tone images were
included in the stock in order to test the efficiency of the
vectorization technique in borderline cases. SCORE was implemented
on an IBM-PC/AT with the intention to install, at a later stage, an
independent workstation in which scanning and vectorization can be
carried out in a single processing step.

Selection Criteria: Test Samples

Drawings of different quality were selected in order to get an idea
of the strong and weak points of the system. The drawings were
characterized by:
- many straight lines
- large hatched areas
- many horizontal and vertical lines
- different line widths
- filled-in areas
- formulas
- diagrams and curves
- handwriting or typewriting
- circuit diagrams (with lines crossing)
    In addition, pictures, hand drawings and drawings with many
textures (bordering on half-tone images) were selected. Drawings
having the above characteristics were classified in three groups
according to complexity. The first group comprised:
- simple and small-size drawings
- medium-size drawings resembling the majority of patent drawings in
  complexity
- full-size format drawings or drawings with many details
    Also investigated was the response to different characteristics
of the master image, e.g.
- pencil drawings
- ink drawings
- black-and-white photographs

Evaluation

The evaluation criteria were:
- accuracy of representation
- vectorization time
- storage requirements
- image quality with high approximation of the algorithm
- suitability for continuous production without manual intervention
- compression factor compared with the raster image
    The tests were successful beyond our expectations. The master
images and the vectorized images were presented to neutral experts

for evaluation; in some cases, the experts found it difficult to
distinguish the master image from the reproduction. The mean
vectorization time on an IBM PC ATO2 was 1.5 minutes, clearly less
than the time required for image scanning. Scanning of the drawings
needs about 6 minutes with a HELL Digigraph 40A40 at a resolution of
16 lines/mm. Storage requirements of the test images amounted, on
average, to 12 KByte. The image quality is excellent, as our
experts confirmed, even with an extremely high degree of polygon run
smoothing. The algorithm was run without failure in all phases of
the experiment. There was no case of blurring, not even in extreme
cases. This proves that our system is reliable, with predictable
quality of reproduction and without any need for human interference,
and suited for application in routine automatic image reproduction.

Representation

The smoothing process produces vectors of different lengths which
are multiples of a unit vector corresponding to the distance between
two pixels. Evaluations have shown that most of the vectors are in
the range of 1 to 50 times the length of the unit vector (see Figure
3). This led us to limit the vector length to a multiple of ± 46 of
the unit vector in both directions (X and Y) and to map the
increments on the 92 characters of the ASCII-Code. As a result, we
now have a text-only database which can be transmitted without
difficulty via normal transmission networks. Text and drawings are
distinguished by control characters transmitted with the document,
which trigger switching from text mode to graphics mode in the
intelligent end user terminal.

Database Implementation and Terminal Support

A pilot version of the patent database with text and drawings has
been implemented in STN. Right now, the bibliographic database is
available at the Karlsruhe node of STN while the drawings have been
implemented in Columbus, Ohio. An integrated solution is scheduled
for the 6.1 version of the STN Messenger retrieval software. STN
has a graphic data structure (GDS) system capable of sending
chemical structure formula in terms of PLOT10 ASCII-Codes. GDS was
modified to enable the transmission of the original image in
portions of closed polygon lines, also represented in ASCII-Codes
and simply differentiated from text or PLOT10 sequence by a control
character.
    There are two possibilities for getting the information wanted,
both text and graphics, from a certain database; by choosing direct
connection to an X.25 port with high transmission velocity and a
protected line protocol of a packet switched network or by selecting
the PAD-version (Packet Assembly Disassembly) with reduced and unprotected line protocols.
    Now that the development phase of the vectorization process is
over, development efforts will focus on the support of different PC
systems to use our service. Any workstation for reception and
visualization of graphic and text data should have the following
components:

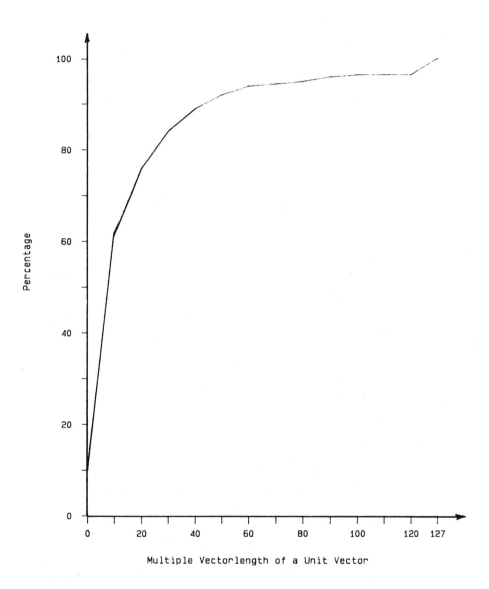

Figure 3. Vector Length Distribution.

- suitable hardware
- communication software
- data management software
- visualization software (polygon filling)
    Owners of an IBM-PC/AT or compatibles with 512K RAM and hard
disk can use FIZ software if their PCs have the following features
added:
- Servonic board for DATEX-P10 connection
- Hercules monochromatic board (720 x 348)
or alternatively:
- serial/parallel adapter with modem (acoutic coupler) for
  connection to DATEX-P (PAD)
- Hercules monochromatic board
    It is planned to provide the customers of STN with a so-called
integrated communication software, capable of fulfilling all
requests concerning the STN dialog components, i.e., searching,
downloading and managing textual data as well as chemical structure,
formula, mathematical symbols and technical drawings.

Data Security Protocol

Errors in polygon chains produce unpredictable lines on the screen
and can destroy the faithful reproduction of a drawing. To protect
image transmission the software provides a controlled line protocol
(KERMIT) slipped over the normal PAD protocol. Such a procedure is
needed only when an unprotected line protocol is applied (for
example the so-called TTY-Protocol supported by a DATEX-P PAD).
Security of transmission costs about 50% more transmission time
depending on the quality of the physical line.

Support of Printers

Visualization of patent drawings via printers has become an
important element of our project, owing to the fact that the high
image quality is best displayed on high-resolution printers. Users
without printers will be offered the possibility of obtaining fast
off-line prints of drawings. The first printer to be supported will
be the Apple LaserWriter. The necessary software is now available.
The development of driver software for the HP Laserjet and the EPSON
LQ 1500 graphics printer will take longer, owing to the fact that
suitable system software is not available or not as powerful as that
of the Apple LaserWriter.
    Support of personal computers with graphics systems will
include the development of new software for generating a metafile
from the vector data which can be interpreted by the PC graphics
packages. So far, there is no general metafile standard, but we
hope that the GKS (8,9) metafile will cover most future metafile
structures. Further plans relate to a special command in STN for
ordering off-line prints of patent drawings.

Outlook

Vectorization of drawings has aroused great interest among CAD/CAM
experts who might profit from automatic conversion to magnetic
storage of technical plans and drawings. Vectorization of drawings
can be applied to all kinds of raster images without half-tones. We
are convinced that our method will open up new prospects in on-line
information supply (10,11).

Literature Cited

1.   Yasuda, Y. Proc. IEEE. 1980, 68 (7), 830-845.
2.   Hunter, R.; Robinson, A. H. Proc. IEEE. 1980, 68 (7), 854-867.
3.   Ackland, B. D.; Weste, N. H. IEEE Trans. Comput. 1981, C30
     (1), 41-48.
4.   Freeman, H. Comput. Surv. 1974, 6 (1), 57-97.
5.   Wall, K.; Danielsson, P.E. Comput. Vision Graphics Image
     Process. 1984, 28 (2), 220-227.
6.   Williams, C. M. Comput. Graphics Image Process 1978, 8, 286-
     293.
7.   Speck, P.T. Thesis, ETH, Zurich, 1984.
8.   Enderle, G.; Kansy, K.; Pfaff, G. Computer Graphics
     Programming. GKS - the Graphic Standard; Springer: Berlin, 1984
9.   Encarnacao, J. Informatik Spektrum 1983, 6 (2).
10. Tittlbach,G. Nachr. Dok. 1986, 37 (4/5), 198-204.
11. Tittlbach,G. Proc. 9th Internat. Online Inf. Mtg., Learned
     Information: Oxford 1985, 95-104.

RECEIVED February 17, 1987

INDEXES

Author Index

Affiliation Index

Subject Index

Production by Cara Aldridge Young
Indexing by Deborah H. Steiner
Jacket design by Carla L. Clemens

Elements typeset by Hot Type Ltd., Washington, DC
Printed and bound by Maple Press Co., York, PA

Recent ACS Books

Personal Computers for Scientists: A Byte at a Time
By Glenn I. Ouchi
288 pp; clothbound; ISBN 0–8412–1001–2

The ACS Style Guide: A Manual for Authors and Editors
Edited by Janet S. Dodd
264 pp; clothbound; ISBN 0–8412–0917–0

Silent Spring Revisited
Edited by Gino J. Marco, Robert M. Hollingworth, and William Durham
214 pp; clothbound; ISBN 0–8412–0980–4

Chemical Demonstrations: A Sourcebook for Teachers
By Lee R. Summerlin and James L. Ealy, Jr.
192 pp; spiral bound; ISBN 0–8412–0923–5

Phosphorus Chemistry in Everyday Living, Second Edition
By Arthur D. F. Toy and Edward N. Walsh
342 pp; clothbound; ISBN 0–8412–1002–0

Pharmacokinetics: Processes and Mathematics
By Peter G. Welling
ACS Monograph 185; 290 pp; ISBN 0–8412–0967–7

Solving Hazardous Waste Problems
Edited by Jurgen H. Exner
ACS Symposium Series 338; 397 pp; ISBN 0–8412–1025–X

Crystallographically Ordered Polymers
Edited by Daniel J. Sandman
ACS Symposium Series 337; 287 pp; ISBN 0–8412–1023–3

Pesticides: Minimizing the Risks
Edited by Nancy N. Ragsdale and Ronald J. Kuhr
ACS Symposium Series 336; 183 pp; ISBN 0–8412–1022–5

Nucleophilicity
Edited by J. Milton Harris and Samuel P. McManus
Advances in Chemistry Series 215; 494 pp; ISBN 0–8412–0952–9

Organic Pollutants in Water
Edited by I. H. Suffet and Murugan Malaiyandi
Advances in Chemistry Series 214; 796 pp; ISBN 0–8412–0951–0

For further information and a free catalog of ACS books, contact:
American Chemical Society
Distribution Office, Department 225
1155 16th Street, NW, Washington, DC 20036
Telephone 800–227–5558